MW00836797

The Nature of Contingency

Quantum Physics as Modal Realism

ALASTAIR WILSON

OXFORD
UNIVERSITY PRESS

OXFORD
UNIVERSITY PRESS

Great Clarendon Street, Oxford, OX2 6DP,
United Kingdom

Oxford University Press is a department of the University of Oxford.
It furthers the University's objective of excellence in research, scholarship,
and education by publishing worldwide. Oxford is a registered trade mark of
Oxford University Press in the UK and in certain other countries

First Edition published in 2020
Impression: 1

Published in the United States of America by Oxford University Press
198 Madison Avenue, New York, NY 10016, United States of America

British Library Cataloguing in Publication Data
Data available

Library of Congress Control Number: 2019945438

ISBN 978–0–19–884621–5

DOI: 10.1093/oso/9780198846215.001.0001

Printed and bound in Great Britain by
Clays Ltd, Elcograf S.p.A.

To Miranda, with love and gratitude for making ours the only world I want to inhabit.

Contents

Preface

This book explores how the conceptual foundations of contemporary physics bear on some traditional metaphysical questions about the nature and structure of objective reality. The specific target of investigation is the metaphysics of modality—contingency, necessity, actuality, chance, and cognate notions—and the specific physical theory that is brought to bear is *Everettian quantum mechanics* (EQM), also known as the *many worlds interpretation*.

EQM is one of the most popular approaches to quantum mechanics amongst theoretical physicists. It is effectively presupposed by a large body of work in quantum cosmology. Unmodified quantum mechanics has passed every empirical test we have been able to devise, and recent experiments have further restricted the range of viable alternative theories by closing loopholes in tests of Bell correlations between entangled quantum systems.[1] In light of progress over the last four decades on the theory of *decoherence*, and of progress over the last two decades in understanding probability in the Everettian setting, EQM—more than ever—appears to be the most natural way to understand contemporary quantum physics. Its potentially radical consequences for metaphysics accordingly deserve examination.

When setting out his metaphysical project of Humean Supervenience, David Lewis memorably rejected the use of arguments from quantum mechanics in metaphysics:

> I am not ready to take lessons in ontology from quantum physics as it now is. First I must see how it looks when it is purified of instrumentalist frivolity, and dares to say something not just about pointer readings but about the constitution of the world; and when it is purified of doublethinking deviant logic; and—most of all—when it is purified of supernatural tales about the power of the observant mind to make things jump.
>
> Lewis (1986a: ix)

[1] I have in mind the 'loophole-free' experiments of Hensen et al. (2015), Giustina et al. (2015), and Shalm et al. (2015).

EQM is how quantum physics looks once it has been purified in all of the ways Lewis demanded, without the addition of any superfluous theoretical structure. It is time to pay attention to its lessons in ontology.

Outside theoretical physics, EQM has hitherto featured mainly as a plot device for science fiction. What has not been appreciated—at least not beyond certain technical debates in philosophy of physics—is the potential of EQM to transform the foundations of metaphysics. In this book I will be posing some perennially difficult metaphysical questions in the Everettian context, and offering some provisional answers to them which make novel use of theoretical resources from quantum physics. The resulting framework—which I call *quantum modal realism*—has strong affinities with the modal realism of David Lewis (Lewis 1986b), but it also has some unique features which set it apart from all extant theories of modality.

The thought that quantum theory might be relevant to the metaphysics of modality is not a new one. 'Quantum logic' interpretations[2] involve profound changes to our understanding of logical consequence; a more radical project in the foundations of metaphysics is hard to imagine. The current project is more conservative: the goal is a minimally revisionary way of incorporating quantum theory into our worldview that leaves untouched our ordinary scientific theorizing about the actual world. In this respect, it is inspired by pioneering work by Simon Saunders (Saunders 1997, 1998) who was the first to make explicit the relevance of EQM to questions asked by metaphysicians about contingency and necessity. Saunders's own views have changed significantly over the two decades since those papers were written. He no longer places such an emphasis on relationality, and is more tolerant of the language of 'many worlds'. In my view, these are steps in the right direction; in this book, I try to take a few more steps.

This idea of this book was conceived while an undergraduate in Oxford in 2002, prompted by tutorials on modality with Bill Newton-Smith and by classes on the philosophy of quantum mechanics with Jeremy Butterfield. I then worked on the project under the guidance of Oliver Pooley, David Wallace, John Hawthorne, Simon Saunders, and Cian Dorr. Numerous friends and colleagues have provided feedback on these ideas along the way, too many to name; I am very grateful to them all. The following deserve special thanks for reading and commenting on substantial chunks of the manuscript: Adam Bales, Chloé de Canson, Christina Conroy, Nina Emery,

[2] The canonical proposal is by Putnam (1968).

Salvatore Florio, David Glick, Dana Goswick, Toby Handfield, Mario Hubert, Noelia Iranzo Ribera, Matthias Jenny, Nicholas Jones, Dan Marshall, Robert Michels, Kristie Miller, John Murphy, Martin Pickup, Mark Pinder, Josh Quirke, Michael Raven, Katie Robertson, Miranda Rose, Alex Silk, Jussi Suikkanen, Tuomas Tahko, Henry Taylor, Paul Tappenden, Naomi Thompson and two referees for Oxford University Press.

This book forms a part of the FraMEPhys project, supported by the European Research Council under the European Union's Horizon 2020 research and innovation programme (grant agreement no. 757295). I also acknowledge financial support from the Arts and Humanities Research Council, from University College, Oxford, from the Australian Research Council, and from a Visiting Fellowship at the Sydney Centre for the Foundations of the Sciences.

Some material from chapters 2 and 3, including figure 3.1, has been reproduced from my previous articles in *The British Journal for the Philosophy of Science*:

'Objective Probability in Everettian Quantum Mechanics' 64(4), December 2013, 709–37

'Everettian Confirmation and Sleeping Beauty' 65(3), September 2014, 573–98

'The Quantum Doomsday Argument' 68(2), June 2017, 597–615

Introduction: Explaining Contingency

0.1 Emergent Contingency

This book argues that quantum theories are best understood as theories about the space of possibilities rather than as theories solely about actuality. When quantum physics is taken seriously in the way first proposed by Hugh Everett III (Everett 1957a), it can offer us direct insight into the metaphysics of possibility, necessity, actuality, chance, and a host of related modal notions. As electromagnetism revealed the nature of light, as acoustics revealed the nature of sound, as statistical mechanics revealed the nature of heat, so quantum physics reveals the nature of contingency. Objective modality is quantum-mechanical.

According to Everettian quantum mechanics (EQM), there exists an enormous plurality of worlds. The entire universe that we see around us, with all its atoms and cities and galaxies, is just one among many universes. Indeed, any way that the laws of quantum physics permit a universe to be is a way in which some universe is. The collection of all of these universes is known as the *Everettian multiverse*. Each universe contained within the multiverse I will call an *Everett world*. You, and all the people you will ever meet, together inhabit just one single Everett world out of the multitude. Although each Everett world is already inconceivably vast, the Everett multiverse is inconceivably vaster.

What the Everettian multiverse is like is not a contingent matter. In a sense to be made precise in section 1.2, there is no possibility that the multiverse could have been any way other than the way it in fact is. Still, there is contingency in the Everettian picture. Indeed, that there should be contingency in EQM is a basic condition on its giving an accurate picture of reality. It is just *obviously true* that things could have been otherwise. I could (and no doubt should) have finished this book years ago; alternatively, it could have been never written at all. You could have been struck by lightning this morning; alternatively, you could have found a winning lottery ticket on the pavement. The Tasmanian tiger could have survived in a small pocket of

The Nature of Contingency: Quantum Physics as Modal Realism. Alastair Wilson, Oxford University Press (2020).

DOI: 10.1093/oso/9780198846215.001.0001

Tarkine forest and have been rediscovered only last month; alternatively, it could have never evolved in the first place. Dinosaurs could have developed human-level intelligence millions of years ago and then laid waste to the Cretaceous environment through rapid industrialization; alternatively, they could have ushered in a scaly era of peace and learning.

Of course, it is possible to deny any of these specific claims about contingent matters. But it is not plausible to deny contingency altogether. Moreover, contingency is an objective feature of the world: it is not simply a function of our representations of the world. What is possible is not just a matter of what we do or do not know, believe, or imagine. But how can genuine contingency be reconciled with the deterministic Everettian multiverse? The answer, implicit in EQM since the very beginning but rarely adequately emphasized, is that contingency relates only to location within the multiverse. What the multiverse is like is non-contingent, but where we are in the multiverse is contingent. Contingency is a wholly self-locating phenomenon, or (to co-opt a term from the theory of semantic content) *essentially indexical.*

One of the most striking features of EQM is that it makes physical contingency into an *emergent* phenomenon. At the fundamental level, the laws of physics are deterministic and physical reality is non-contingent. Only at non-fundamental (derivative) levels do we find indeterministic physical laws and genuine contingency in nature. This is a deeply unfamiliar picture; can we make sense of it? I think that we can. The history of science contains numerous cases in which some phenomenon, once assumed to be basic and irreducible, has been discovered to be—somehow or other—emergent. Earth, air, fire, water, life, mentality, solidity, colour: all of these have turned out not to be basic ingredients of the world but instead to be manifestations of previously unsuspected kinds of organization at more fundamental levels. To take an example a little closer to the present case, recent work in quantum gravity has begun to take very seriously emergence hypotheses concerning space and time. Still, even in the company of these radical hypotheses, the idea that contingency itself is emergent stands out as especially radical.

The theoretical role of chance is to provide an objective measure over a space of physically possible histories. Since physical contingency is emergent in EQM, so too is chance. This Everettian vision of chance as emergent is becoming familiar through the work of Wallace (2012), Saunders (2005, 2010b) and others; but it naturally leads on to an unfamiliar treatment of modality as a whole. In chapter 1, I argue that we have overwhelming reason to extend the Everettian analysis of physical contingency to an analysis of contingency more generally. I introduce and defend the principle *Alignment*, which tells us that to be a metaphysically possible world is to be an Everett world. Alignment forms the core of my new naturalistic theory of modality: *quantum modal realism*.

The remainder of chapter 1 introduces the main features of quantum modal realism, and catalogues the various theoretical benefits within metaphysics to which quantum modal realists may lay claim. Quantum modal realism provides for a comprehensive account of qualitative—or *de dicto*—modality, giving us everything we need in order to say how things in general could have been, would have been, will probably be, cause other things to be, seem to be, are acted upon, and so forth. By employing the key Lewisian innovation of *counterpart theory*, quantum modal realists may extend their treatment from qualitative modality to non-qualitative—or *de re*—modality, which concerns how particular things like you or me could or would be. Quantum modal realism underwrites an orthodox S5 modal logic. The reductive account of laws and chances that it provides allows for some strikingly simple realist theories of the semantics for counterfactual discourse, and thus for robust counterfactual-based theories of causation. It also provides resources for analysing semantic content, both coarse-grained and fine-grained, and for characterizing the nature of properties. Quantum modal realists thus inherit most of the riches of what David Lewis called the 'philosophers' paradise' of his own modal realism (Lewis 1986b). Moreover, the Everettian multiverse comes equipped with a global objective chance measure which fills crucial gaps in the Lewisian story; this chance measure permits explanations of the knowability and rational relevance of modal facts which Lewisian modal realism cannot support.

Chapter 2 delves into the Everettian approach to quantum mechanics that is at the core of quantum modal realism, sketching and motivating the physical and metaphysical components of the quantum modal realist account of modality. After a brief summary of how EQM resolves the notorious quantum measurement problem, I show how some core principles of quantum modal realism that connect the physics of EQM with the metaphysics of modality play a key role in resolving long-standing problems with probability in the Everettian setting. These principles are fleshed out in chapter 3 into a full quantum modal realist theory of objective chance, including an Everettian Principal Principle connecting chance and credence. While chance is emergent in the Everettian picture, it is nonetheless a demonstrably excellent candidate to play the theoretical role of objective chance: it is that element of physical reality to which rational epistemic agents embedded in Everett worlds conform their expectations.

What goes for the Everettian multiverse and for its constituent worlds goes also for the physical laws which govern them. Although the laws of the multiverse—the fundamental laws of quantum mechanics—are deterministic and non-contingent, quantum modal realists can also make clear sense of derivative laws governing events within individual worlds. Such laws may be indeterministic and/or contingent. Chapter 4 takes up the task of sorting out the different ways in which the ideology of lawhood may be applied within EQM, and in the process offers a new and attractive theory of laws of nature that reconciles the prominent Humean and anti-Humean viewpoints in a previously unanticipated way.

While the quantum modal realist's fundamental reality remains fully determinate, the emergent multiverse of EQM is indeterminate in two characteristic ways. Everett worlds are indeterminate both in number and in nature. In chapter 5, I model this indeterminacy using the precisificational approach widespread in recent metaphysical discussions. I argue that while indeterminacy in both the nature and the number of Everett worlds is ultimately semantic or epistemic in origin, the indeterminacy in the nature of these worlds may at the same time be regarded as (a naturalized and emergent form of) metaphysical indeterminacy.

A book like this must necessarily make a number of controversial assumptions. The most significant of these assumptions concerns physics rather than philosophy: in what follows, I shall be assuming without substantial argument that EQM is a coherent and defensible framework for understanding quantum phenomena. Although some details of the formulation of EQM will be taken up in chapters 2 and 3, in general I shall be relying on the work of others in explicating and defending the Everettian approach. While I think the project of this book, if successful, is capable of providing some limited independent evidential support for EQM, I shall not attempt to make the case for that claim here.

It is interesting to ask whether any other approaches to quantum theory can support a theory of modality of the general kind that I propose. Some elements of my approach can certainly be emulated within Bohmian mechanics or within dynamical collapse theories. However, any explanation of quantum phenomena that posits a fundamental indeterministic or stochastic collapse mechanism will be unable to capture what I see as the central advantage of quantum modal realism—its genuinely reductive character. No-collapse approaches such as Bohmian mechanics may also not turn out to be reductive, if they retain non-quantum contingency in matters such as the initial distribution of particles. (A modal realist version of the Bohmian approach seems potentially viable,[1] but it would start to closely resemble EQM.) In any case, the Everettian approach provides the most obvious route to a reductive quantum-mechanical account of modality, and this is the route I will be exploring in this book.

In section 0.2, I turn to the target of my reductive project: objective modality. Philosophical readers who are already sufficiently puzzled by the nature of modality to see it as standing in need of explanation may skip to section 0.3. Those wondering what modality is, or why it should be especially hard to accommodate in a scientific worldview, should read on.

0.2 Theories of Modality

Ordinary discourse, as well as scientific discourse, is littered with apparent modal truths: explicit claims of possibility or necessity, counterfactuals, and ascriptions of abilities or dispositions to name but a few. As Sellars (1948, 1957) emphasized, much of our linguistic practice bears disguised modal commitments. In light of this, it is an embarrassment to metaphysics that it has been unable as yet to produce any plausible candidate account of the nature and source of modality. No extant theory is credible.

Early twentieth-century accounts of modality emphasized the role of convention in explaining modal discourse. In their simplest form, such accounts identified the possible with that which is not ruled out by convention. Simple conventionalism eats its own tail; it cannot give a sensible account of which conventions are themselves possible and which impossible. Several authors, recognizing the limits of conventionalism, looked for accounts of modality with an objective element. The natural place to start was with objective features of language. Carnap bequeathed us the notion of a state-description (Carnap 1946), which—while adequate for various purposes in formal semantics—provides little insight into the nature of modality. Linguistic accounts of modality invariably need to presuppose modal notions in the form of a primitive notion of consistency of sentences or propositions. Modality itself goes unexplained.

[1] In conversation, Sheldon Goldstein has expressed some sympathy for a Bohmian modal realist approach.

Quine's notorious hostility to modality is often overstated; his primary objection was to *de re* modal properties possessed by individual objects, which he saw as indefensibly privileging one way of picking out an object over other ways of picking out exactly the same object. Indeed the 'metaphysical turn' that Quine triggered (intentionally or not) within analytic philosophy led to a renewed interest in the nature of modality. Kripke's formal results in model-theoretic semantics for modal logic assuaged any fears about incoherence or inconsistency of the main logical systems, and his informal *Naming and Necessity* (Kripke 1980) soothed Quinean scruples about *de re* modality by providing a compelling intuitive gloss on reasoning about possible individuals via *possible worlds*.

Possible worlds provide a compelling and fertile set of tools for formalizing our thinking about modality, but they only sharpen the problem of accounting for the metaphysical basis of modal thought and talk. If worlds are to be more than a bare formal device for model-theoretic metalogical investigations into modal logic then we must give them some positive metaphysical interpretation (see Williamson 2013: 139–40 for discussion). Lewis put the point memorably:

> For [metalogical results], we need no possible worlds. We need sets of entities which, for heuristic guidance, 'may be regarded as' possible worlds, but which in truth may be anything you please. We are doing mathematics, not metaphysics. Where we need possible worlds, rather, is in applying the results of these metalogical investigations... To apply the results you have to incur a commitment to some substantive analysis of modality.
>
> Lewis (1986b: 17–19)

If we are to understand modality in terms of possible worlds, we cannot avoid the core metaphysical questions: what are possible worlds, and what sorts of possible worlds are there? Subsequent work has helped to identify two crucial constraints on adequate answers to these questions: the answers must help us to make sense of how we know which worlds are possible (the *epistemic challenge*) and they must help us to make sense of how possible worlds are relevant to our practical interests (the *practical challenge*). How can we know about, and why should we care about, what can be the case as opposed to what is the case? In the remainder of this section I argue that all extant theories of the nature of possible worlds fail to address one or both of these challenges. This book as a whole makes the case that quantum modal realism has the resources to meet both challenges.

Treating possible worlds as linguistic entities amounts to adopting the Carnapian state-description approach as a metaphysical hypothesis rather than simply as a formal model. Linguistic theories of possible worlds do passably well with respect to the epistemic challenge: competence with a language at least in principle gives us the ability to detect which strings of that language are consistent. However, linguistic theories seem to render the practical problem insoluble. What difference between consistent and inconsistent sets of sentences explains why the former but not the latter are relevant to our practical interests? The obvious answer—that the members of consistent sets of sentences could all be jointly true, while the members of inconsistent sets of sentences could not all be jointly true—is unavailable to those who seek to analyse modality itself in linguistic terms.

As part of a general move away from linguistic theorizing towards metaphysical theorizing, Alvin Plantinga (1974) proposed replacing state-descriptions with states of affairs: worlds are a complex kind of property rather than anything linguistic. This change, however, achieves little with respect to the circularity problem: it simply turns a primitive notion of consistency of predicates into a primitive notion of compossibility of properties. We have no grip on the notion of compossibility of two properties unless it amounts to it being possible that something should have both properties; indeed, it seems exactly backwards to analyse possibility in terms of compossibility. When measured against the epistemic and practical challenges, the property-based theory is of no use at all. To be told that a world is possible just in case it does not instantiate incompossible properties is no help unless we antecedently know how to tell whether two properties are incompossible; and without an account of incompossibility in independent terms, it is unclear why two properties being incompossible should lead us to disregard in our practical reasoning any scenario which combines those properties.

One way of avoiding the pitfall of accounting for possibility in terms of some notion that implicitly presupposes possibility is to refuse to account for possibility in independent terms at all. In this spirit, Peter van Inwagen proposed an account of possible worlds as *sui generis* abstract entities, as points in an abstract space about which nothing could be said except that they represent certain states of affairs and fail to represent other states of affairs. This account, dubbed 'magical' and criticized as mysterious by Lewis (1986b), was defended by van Inwagen (1986) on grounds of good company; he argued that any methodological difficulty with a magical account of possible worlds was also shared by the realism about set theory endorsed by Lewis. We need not dwell on the comparison with set theory, since our purpose is to assess how well a theory of modality measures up against the epistemic and practical challenges.

The magical proposal fares at least as badly as any account previously considered. Since it is part of the account that possible worlds have no features in virtue of which they represent something as possible, they cannot—as a matter of principle—have any features that would explain how we can know about them or that could explain how knowledge of them would be relevant to our practical interests.

The circularity at the heart of the linguistic, property-based and magical accounts is deeply unsatisfactory. To explain how possible worlds can play their allotted theoretical roles, defenders of these theories need to appeal to the very modal judgements that possible worlds are meant to underwrite. Perhaps we could reluctantly conclude that this is a kind of explanatory circle we have to learn to live with. Every metaphysical system must contain some unanalysed primitives, we might argue; and why should these primitives not be modal ones? But even if we endorse this line of thought, we might still be dissatisfied with the arcane metaphysics often smuggled in under the title 'possible worlds' or 'states of affairs'. For Plantinga, states of affairs are *sui generis* abstract objects; and his theory invokes a multitude of abstract uninstantiated individual essences. Van Inwagen likewise adopts a view wherein possible worlds have no intrinsic qualitative features to distinguish them but are merely abstract placeholders in a system of abstract relations. Each would have us believe in a host of *sui generis* abstract objects, the essence of each of which is to represent some complete possible world and which plays that role despite lacking any internal qualitative structure. One does not need to be an austere Quinean to worry about this proliferation of heavy-duty metaphysics.

A striking alternative to the positing of exotic candidates for possible worlds is the *modal realism* of David Lewis (1986b). While the Lewisian view is radically liberal with respect to the quantity of concrete entities it endorses, it is radically conservative with respect to their qualities. Lewis's proposal for the nature of possible worlds is disarmingly simple: they are things just like the actual world that we inhabit, just more of the same old goings-on. Nothing could be more familiar than the material world in which we live and move and have our being.[2] But despite the qualitative parsimony

[2] This familiarity is undermined somewhat by Lewis's acknowledgement of worlds including very peculiar entities: ghosts, 'epiphenomenal rubbish', and the like. But quantum modal realists ought not, indeed cannot, recognize worlds like those.

it involves, almost nobody apart from Lewis has ever endorsed Lewisian modal realism. While the reasons offered for rejecting it have been many and various, to my mind the most potent objections to modal realism are the same challenges that I have already argued are fatal to the linguistic, property-based and magical accounts. That is, Lewisian modal realism is unable to account for our envisaged epistemic access to the plurality of worlds, and it is unable to account for their envisaged relevance to our practical interests. Since my own quantum modal realism has more in common with Lewisian modal realism than with any of the views so far discussed, it will be worth spelling out these challenges in more detail to see how they apply specifically to the Lewisian proposal.

The epistemic challenge is one that Lewis considered explicitly.[3] He cites Skyrms (1980: 326), who puts the point in causal terms: how can we have knowledge of other Lewisian worlds when we are causally isolated from them? Still, Lewis never satisfactorily addressed the challenge. Indeed, the structure of his response to the epistemic objection has no hope of succeeding unsupplemented: it presupposes exactly what the challenge requires it to explain. Here is the response: 'If modal knowledge is what I say it is, and if we have the modal knowledge that we think we do, then we have abundant knowledge of the existence of concrete individuals not causally related to us in any way' (Lewis 1986b: 111). And:

> [H]ow do we come by the modal opinions that we in fact hold? . . . In the mathematical case, the answer is that we come by our opinions largely by reasoning from general principles that we already accept . . . I suppose the answer in the modal case is similar. I think our everyday modal opinions are, in large measure, consequences of a principle of recombination . . .
>
> Lewis (1986b: 113)

The epistemological picture that Lewis is sketching is as follows (I paraphrase): if modal realism is correct, then our modal beliefs just are beliefs about the plurality of worlds. So, if modal realism is correct, then we do in fact have numerous true beliefs about the plurality of worlds, and it cannot be the case that it is impossible to acquire true beliefs about the plurality of worlds. Moreover, if these beliefs are in fact true, they are necessarily true since they concern non-contingent matters. If the method used to form

[3] The challenge may originate with Richards (1975).

them could not easily have led to the formation of different beliefs, then our true beliefs might even qualify as knowledge.

The problem with this overall epistemological picture is that it fails to identify any evidence we have that counts in favour of the principle of recombination. Accordingly it fails to explain how and why the entire epistemic practice of forming beliefs about possible worlds through judgements based on recombination got started. Consider an intelligent species with no modal knowledge: they know some facts about their actual world, but they do not have any modal beliefs at all, and they do not suspect that any concrete worlds exist other than their own. Such a species would have no reason to believe that a principle of recombination holds. The key question which Lewis cannot answer is: how could we have got here, with our extensive true modal beliefs, from there, where we had none? It is one thing to mount a defensive operation of some territory that you control, but quite another to conquer the territory in the first place. To conquer epistemic territory, you need evidence—and Lewis offers us none.

It is instructive to compare my argument here with Ross Cameron's recent claim (Cameron 2007a) that we cannot be justified in believing Lewisian modal realism. Like me, Cameron criticizes Lewis for mounting a purely defensive operation, but the basis for his criticism is that Lewis is not entitled to presuppose the metaphysics of modal realism when providing an epistemological story about our knowledge of individual modal truth. According to Cameron, '[w]hat Lewis needs to do to respond to the epistemological objection is to provide reason *independent* of the truth of Lewisian realism to think that one can have knowledge about non-actual possible worlds' (Cameron 2007a: 149). I think this requirement is much too strong: nobody would think it a reasonable requirement of an atomist that they be able to provide reason *independent of the truth of the atomic theory of matter* to think that we can have knowledge of atoms and their properties. All of our knowledge of atoms, at least initially, came from inferences from the behaviour of the macroscopic world (Brownian motion, for example); and these inferences only in fact provide knowledge of atoms and their properties if the atomic theory of matter is in fact correct.

Unlike Cameron's objection, my version of the epistemological objection to modal realism does not problematically overgeneralize. It *is* incumbent on an atomist to explain how (if atomism is correct) we came to have knowledge of the kinds of phenomena that are (by their own lights) dependent on the existence of atoms and their properties. How (if atomism is correct) did we first acquire knowledge that there is Brownian motion going on when pollen grains are suspended in water? An answer, of course, is that we ourselves are made of atoms which causally interact with atoms within the water via a causal chain involving a microscope. The Lewisian realist, however, cannot offer any comparable story here. Other Lewisian possible worlds bear no constitutive, causal, or other explanatory relations to the observable goings-on within our own world. If Lewisian modal realism is correct, then how we ascended to our current state of modal knowledge is an intractable mystery, even if our current modal beliefs were (inexplicably) formed de facto reliably. This is my favoured version of the epistemic challenge to Lewisian modal realism.

The practical challenge to Lewisian modal realism has been most forcefully raised by Jubien (1988). Lewis's response to the practical challenge closely mirrors his response to the epistemic challenge, and it fails for a closely related reason; hence, we can be briefer here. Lewis correctly notes that our modal knowledge is of great practical relevance to us (never mind exactly why or how); he points out that if modal realism is correct, then our modal knowledge is knowledge about a plurality of worlds; he infers that if modal realism is correct, then knowledge about a plurality of worlds is of great practical relevance to us; finally, he pivots on this inference, concluding that any general principle about the practical irrelevance of entities which are causally isolated from us must be false. Again, the problem with this line of thought is that it constitutes a defensive operation where what is needed is a positive account.

Consider a species that does not perceive any relevance to their own practical interests of bizarre hypotheses about a plurality of spatiotemporally isolated concrete worlds. These creatures care about the nature of the world they inhabit, but they could not care less about other causally isolated worlds. Where are such creatures going wrong?—and what could we possibly say to induce them to take an interest in the other Lewisian worlds? As long as Lewis has no good answer to these questions, he has no satisfactory response to the practical challenge.

Despite the compelling epistemic and practical challenges to Lewisian modal realism, there remains something very attractive about it. It is an elegant—indeed, a beautiful—theory. While quantitatively unparsimonious, it is qualitatively parsimonious: there are many worlds, but each of them is familiar from our actual world. And while hard to believe, it is very easy to grasp. It is therefore unsurprising that various parasitic theories have been advanced which borrow from the explanatory resources of Lewisian modal realism while distancing themselves from Lewisian modal realism's more problematic commitments. The most well known of these theories is modal fictionalism (Rosen 1990) which cashes out modal discourse via the pretence that a Lewisian plurality of worlds exists. A relative of modal fictionalism is modal agnosticism (Divers 2004), which makes use of the Lewisian modal realist story to understand modal discourse while remaining neutral on whether the Lewisian plurality is real or fictional.

While fictionalism and agnosticism may have advantages with respect to the epistemic challenges—if we are collective authors of the fiction, then perhaps we have collective authorial privilege over its content?—these parasitic theories do not help with respect to the practical challenge. If the Lewisian story is false, then why have we developed an ingrained practice of reasoning under the pretence that it is correct? Or, if the Lewisian story is something on which we should be rationally agnostic, then why have we adopted a deep-rooted practice of reasoning conditional on the supposition that it is correct? In both cases, the heuristic of the amodal species introduced above helps to underscore the point. There seems to be no way to rationally induce a species interested only in the goings-on within their actual world to take an interest in a fiction according to which there are many other causally isolated worlds, and no motivation for members of such a species to dwell on what follows from the supposition that such a plurality exists.

The general consensus amongst metaphysicians is that Lewisian modal realism is a bizarre and extravagant fairy tale, and ultimately not a hypothesis worth taking seriously. The story of the Lewisian plurality is one which we could never be in an epistemic position to substantiate and one which could never be relevant to any of our practical interests. And with this rejection of Lewisian modal realism the debate has largely stalled. None of the theories on the table are capable—even in principle—of explaining why we modalize.[4]

I do not take the above arguments, quickly sketched as they are, to be decisive; nor shall I dwell any longer on the claim that all extant theories of modality are untenable. My main thesis in this book is a more positive one: that EQM allows us to formulate a powerful and tenable reductive theory. What follows is an investigation into how the quantum modal realist position should best be formulated, with the hope that the explanatory benefits will speak for themselves.

So much for the current state of the metaphysics of modality. I turn in the next section to an important but under-appreciated question in the area: the

[4] Divers (2010) argues that this challenge is urgent and underappreciated.

question of whether the primary target of our modal theorizing should be necessity or contingency.

0.3 Necessity or Contingency First?

The majority of systematic treatments of modal metaphysics take necessity as their primary explanatory target. The general thought, not often articulated, seems to be that contingency is the 'default' modal status for a proposition and that the job of a theory of modality is to provide an account of necessity as a deviation from this default status. Gideon Rosen captures this outlook nicely as follows:

> [M]etaphysical possibility is, as it were, the default status for propositions. When the question arises, 'Is P metaphysically possible?' the first question we ask is 'Why shouldn't it be possible?'... P is metaphysically possible unless there is some reason why it should not be—unless there is, as we say, some sort of obstacle to its possibility. Moreover, the only such obstacle we recognize is latent absurdity or contradiction. Rosen (2006: 23)

We may take as an example Kit Fine's (1994) essentialist account of modality, according to which (roughly) a proposition is necessary just in case it is true in virtue of the essences of some things. The operative presumption seems to be that, once such a theory of modality has given us a story about the nature of necessity, a story about contingency comes for free. If the essences of things are silent on whether a proposition or its negation holds, then Finean essentialism counts the proposition as contingent.

Lewisian modal realism is distinctive in that it inverts the above conception of the target of a theory of modality. There is a clear sense in which the basic ontological claims of modal realism—those claims needed to describe the whole plurality of worlds—are necessary. (See section 1.2 for further elaboration of this point, and a discussion of the connection to modal logic.) Since Lewisian modal realism is intended to be reductive, these ontological claims need to be *conceptually non-modal*, in the sense of Divers & Melia (2002): modal realists must not presuppose any modal notions in formulating their theory, if it is to be genuinely reductive. Accordingly, modal realists must not account for modality in terms of any claims whose modal status itself calls for explanation. In this sense, then, modal realists must take necessity as the default status for a truth, and contingency as the modally loaded status for which they provide a substantive analysis. Since the version of quantum modal realism that I will spell out in this chapter and the next is a close relative of Lewisian modal realism, quantum modal realism too is best understood as making necessity prior to contingency.

We have, then, a distinction between what I will call *necessity-first* and *contingency-first* approaches to analysing modality. These reflect different attitudes to what is puzzling about modality, and hence different adequacy conditions on a theory of modality. The distinction between necessity-first and contingency-first approaches pertains to the metaphysical order of priority between the two phenomena, not to the epistemological order of priority. It may be, for example, that our knowledge or understanding of contingent propositions (I'm hungry; that plant is edible; tigers are dangerous; it is raining) is a necessary precondition of our having knowledge or understanding of necessary propositions. This does not refute the necessity-first approach, which is a view about the order of metaphysical priority between necessity and contingency, just as the epistemological priority of facts about colours over facts about wavelengths of light does not refute the view that colours can be reduced to wavelengths.

Co-opting the language of grounding that is widely used in contemporary metaphysics[5] we may formulate the contrast as follows:

Necessity-first: Necessary facts are such that nothing grounds their necessity.
Contingency-first: Contingent facts are such that nothing grounds their contingency.

Lewisian modal realism is clearly a necessity-first theory by this classification: the necessary facts about the plurality of worlds ground the contingency of those facts that are contingent, but nothing grounds their own necessity. For Lewis, it is because the principle of recombination holds that there is a world containing a counterpart of me writing a longer book, and hence the (necessary) principle of recombination grounds that it is contingent how long this book is. But for facts about the plurality to be necessary just is for them to obtain (see section 1.2 for further discussion), and hence the necessity of these facts is not grounded in anything.

[5] Readers may give the grounding terminology a thin reading, as just marking direction of explanation in metaphysics. For further discussion of the role of grounding in my project, see section 0.4 of the introduction.

These formulations of the necessity-first and contingency-first approaches make them neither exclusive nor exhaustive: it is conceivable that some unwieldy theory might treat both necessity and contingency as distinct primitives, or even that an ingenious way to reject both principles might be found. Perhaps Sider's neo-conventionalism about modality (Sider 2000, 2011) would be a view of this kind, since it regards both necessity and contingency as grounded in our modal conventions. Be that as it may, the quantum modal realism that I develop in this book involves accepting Necessity-first and rejecting Contingency-first.

The distinction between contingency-first and necessity-first approaches to modality has not attracted much attention in the metametaphysics literature; this is surprising, because it has some immediate and striking methodological consequences. When metaphysicians disagree about whether some scenario is metaphysically possible, it is typically thought that the burden of proof is on those who would regard it as impossible to explain what makes it impossible. This fits with the contingency-first outlook; if the default status for a proposition is contingency, then the proposition that the disputed scenario obtains is presumptively possible. In a slogan, things are possible unless proven otherwise.

The contingency-first approach puts the onus on views according to which some claims are necessary to give a non-trivial account of their necessity. In this vein we find Ross Cameron (2007b) defending the contingency of principles about composition by encumbering his opponent with the burden of proof; from the contingency-first perspective, it is quite reasonable to ask for some distinctive feature of claims about composition which overrides the default status of contingency. I think it is not generally recognized just how powerful this burden-shifting manoeuvre can be; the contingency-first perspective makes it very hard to sustain any synthetic necessities whatsoever. Logical inconsistency of a proposition, or of its negation, is a clear defeater for the proposition's presumptive status of contingency; it is controversial what else, if anything, is able to defeat this status. By inverting the contingency-first picture, Lewisian modal realism and quantum modal realism give rise to a radical shift in the dialectical position. Now the default modal status for any factual claim is non-contingency; if we want to establish that some proposition is contingent, it is incumbent on us to show that it falls into the special class of contingent propositions. Unless some claim is essentially indexical—unless it is a matter of self-location within the Everettian multiverse—then it has its truth value non-contingently.

Adherence to the necessity-first picture provides a route for modal realists, of Lewisian and Everettian stripes alike, to dodge Simon Blackburn's dilemma for theories of the 'source of necessity' (Blackburn 1986). According to Blackburn, a theory must either account for the source of necessity in terms of contingent truths, in which case 'the original necessity has not been explained or identified, so much as undermined' (ibid.: 53), or else they must account for the source of necessity in terms of necessary truths, in which case 'there will be the same bad residual "must"' (ibid.: 53). The dilemma is that necessity itself cannot be explained: '[e]ither the explanandum shares the modal status of the original, and leaves us dissatisfied, or it does not, and leaves us equally dissatisfied' (ibid.: 53).

Responses to the dilemma have varied (Hale 2002, Lange 2008, Cameron 2010) but no response has been widely accepted, and the dilemma stands as a major challenge to reductive theories of modality. The necessity-first picture dissolves the dilemma in a novel way, since it reconceives the task of a theory of the source of modality as the task of providing an explanatory account of contingency. When the conception of the required explanatory task is inverted in this way, so must Blackburn's dilemma be inverted; and, as a challenge to theories of the source of contingency, the inverted dilemma is uncompelling. In a sense, necessity itself does go unexplained on the quantum modal realist picture. From the necessity-first perspective, though, this is exactly as it should be. Rather than offering some independent positive account of the status of necessity, necessity-first approaches are in the business of dissolving the need for any such positive account.

What modal realists do, against a theoretical background of purely necessary claims, is to locate an essentially indexical subject matter as the source of contingency. The role of necessities in the modal realist approach is collectively to *make room* for contingency, not to entail any particular contingent truths. There is no residual mystery concerning how indexical questions remain open once the non-indexical facts have all been fixed. The contingency of all contingent truths can in this way be explained wholly in terms of necessary truths.

0.4 Naturalistic Metaphysics

This book aims to draw metaphysical conclusions from considerations of quantum physics. There are familiar obstacles to this general approach. A single physical theory can bear a variety of legitimate interpretations that differ metaphysically from one another, so even if we had a complete and correct physics it would not by itself answer any of the central questions of metaphysics for us. It will not do to respond to this challenge by advocating full deference to the metaphysical opinions of actual scientists. An understanding of how a theory is applied certainly helps with avoiding confused or unworkable interpretations of that theory. But there are limits: we should distinguish between metaphysical assumptions that scientists make for good scientific reasons, and assumptions that they make as a result of some irrelevant historical or psychological contingency. Observing this contrast requires a more nuanced approach to letting physics guide metaphysics.

How is it possible for physics to bear on metaphysics without supplanting it completely? The methodological position I want to endorse has three main components:

1) *Interpretational Metasemantics*: the meanings of a linguistic community's terms are just the meanings that are assigned to them by the best theory of that community's linguistic behaviour.
2) *Confirmational Holism*: theories cannot be adequately assessed in isolation, and the ultimate unit of theoretical comparison is the complete system-of-the-world.
3) *Physics/Metaphysics Evidential Asymmetry*: central principles of physics are better confirmed than central principles of metaphysics, so modifying metaphysics is—*ceteris paribus*—less of a theoretical cost than modifying physics.

Interpretational metasemantics as I employ it here is a Lewisian doctrine, with roots in Quine and Davidson.[6] Confirmational holism is primarily associated with Quine within analytic philosophy (though it has echoes of Neurath's boat, and of earlier pragmatists).[7] The roots of the physics/metaphysics evidential asymmetry are in the scientistic spirit of logical positivism and logical empiricism, transplanted into contemporary metaphysics through the influence of Quine, Smart and Dennett.[8] In recognition of the common factor in the origin of these theses, I will refer to the conjunction of

[6] See, especially, Lewis (1974, 1983b), Davidson (1973), Quine (1960a). 'Best theory' may evidently be cashed out in various ways; I will aim to remain neutral on the precise form that interpretational metasemantics should take.

[7] See, especially, Quine (1951, 1957), Neurath (1944), Duhem (1906).

[8] See, especially, Quine (1969b), Smart (1963), and Dennett (1969).

1–3 as the *Quinean methodology* for metaphysics. What work is done by the various elements of the Quinean methodology?

Firstly, interpretational metasemantics reassures us that there is no gap between a theory about the meaning of our metaphysical terms making the best overall sense of our usage of those terms and that theory being correct. If linguistic usage partly determines meaning through a best-interpretation competition, we avoid the risks that our metaphysical terms systematically fail to refer to elements of reality or that we are systematically mistaken about the properties of their referents. Interpretational metasemantics thus blunts the force of sceptical arguments in metaphysics, construed as arguments from *global* underdetermination of theory by data.

Secondly, confirmational holism is a key ingredient in the resolution of metaphysical questions in the face of the more interesting *local* form of underdetermination argument. Such arguments highlight structural similarities between different metaphysical accounts of some specific notion, threatening to make the choice between these accounts into an arbitrary one. However, bringing in aspects of the wider theoretical context allows us to break these local cases of underdetermination: an example of this procedure, as it applies to the mereological structure of Everett worlds, features in chapters 2 and 3.

Thirdly, the Quinean's asymmetric attitude to physics and metaphysics reflects the vast disparity in the evidence we have in the two domains. Fundamental physics has astounding and systematic empirical evidential support. Fundamental metaphysics has no such empirical evidential basis; and indeed, sometimes it is hard to see what evidence we could possibly bring to bear on its questions. Bennett (2009) argues that in the face of this evidential drought we should typically suspend judgement on matters metaphysical. Sometimes, though, indirect evidence is good enough; and if a metaphysical principle is indispensable to a total theory that makes best sense of known physics, that source of evidence for the principle may be the best we can have. Being so well-confirmed empirically, our physical theories should be meddled with as little as possible in formulating total theory. Instead of being modified or rejected on metaphysical grounds, physical theories should be supplemented with whatever metaphysics suits them best. In other words, we should compare complete physics-plus-metaphysics package deals, and adopt whichever metaphysical principles are part of the best package.

Applications of the Quinean methodology are now familiar from the work of Lewis, Jackson, Menzies, Sider and others; they sometimes go under the banner of the 'Canberra plan'.[9] Mental states, objective moral value, causal relations, dispositions, and other metaphysicalia have been located in naturalistic worldviews by Canberra planners. Even if our patterns of use of metaphysical terms do not match up perfectly with anything fundamental within the worldview of physics, such terms may still be taken to successfully refer if enough of their use can be preserved by identifying them with some non-fundamental aspect of reality. Latitude with respect to this use can be relatively wide: there need be no great cost to demoting individual platitudes about metaphysical terms to false beliefs, whereas there is a substantial empirical cost to discarding central principles of contemporary physics.

The version of naturalistic metaphysics I am advocating may be understood as an inter-theoretic reduction of problematic metaphysics to unproblematic physics, with interpretational meta-semantics providing a global assurance that the best available reduction is *ipso facto* correct. But do not be alarmed by the word 'reduction'; the reductionism involved is of a very flexible sort. What I have in mind is (generalized) Nagelian reduction, the programme of explaining the success of a reduced theory by constructing an analogue of the reduced theory using the resources of the reducing theory.[10] Ernest Nagel (1961) gives a characterization of theoretical unification which relies neither on the mereological assumptions of Oppenheim and Putnam (1958) nor on the bridge laws present in Nagel's earlier proposals and later influentially criticized by Fodor (1974). The key difference is that a Nagelian reduction is reduction of theories, and not reduction of theoretical entities: they consist in formal explanations of why the reduced theory obtains in terms of the reducing theory. The explanation can take many different forms depending on the forms of the theories in question. What is important in Nagelian reduction is an explanatory connection between the theories; in particular there is no requirement that the entities of the reduced theory be composed mereologically out of the entities of the reducing theory.

The analogy with Nagelian reduction helps us zero in on the cleanest way of formulating metaphysical claims. By analogy with the theoretical

[9] On the Canberra Plan, see Nolan (1996), Jackson (1998), and Braddon-Mitchell & Nola (2009).

[10] These reductions will typically require 'additional assumptions' which correspond to conditions under which the reduced theory applies. Nagel's central example is the deduction of the Boyle/Charles law of gases from kinetic theory; the 'additional assumptions' in this case are modelling conditions for gases: treating the particles as elastic spheres, assuming a large number of particles, and so on.

identifications that emerge from particular Nagelian reductions (heat is molecular motion; lightning is electrical discharge; water is H_2O; etc.), naturalistic metaphysics can be cast as the project of finding explanatory accounts of the individual terms appearing in distinctively metaphysical questions. One familiar form such identifications may take is 'to be an X is to be a Y', but the more general form is propositional: 'for it to be the case that X is for it to be the case that Y'. Such *metaphysical analyses* entail necessary and sufficient conditions for the application of the term being analysed.[11]

The main advantage of stating metaphysical doctrines in terms of metaphysical analyses is that it maps onto a characteristic form of explanation in the natural sciences: *reductive explanation*. In textbook cases of reduction, such as the reduction of thermodynamics to statistical mechanics, one component of the explanation relies on identities between the referents of certain theoretical terms. For example, in standard accounts of the relationship between thermodynamics and statistical mechanics, the thermodynamic quantity of temperature is identified with the statistical-mechanical quantity of mean molecular kinetic energy.[12] This identification is explanatory because it forestalls certain 'why'-questions which one might initially want to ask; for example, once we recognize that temperature just is mean molecular kinetic energy, there is no need to explain why two gases which match in mean molecular kinetic energy also match in temperature.[13]

This kind of neutralizing of explanatory demands through theoretical identification is distinctive of the analysis-based formulation of metaphysical doctrines. There is no mystery about why rapid oxidation always accompanies fire, or about why electrical discharge occurs wherever lightning does, or about why water is invariably co-located with hydrogen dioxide, or (more prosaically) about why bachelors are without exception unmarried. In each case, some metaphysical analysis is serving to reduce the space of distinct possibilities for the world; if we know that to be an A is to be

[11] When we make any metaphysical claim of this sort, we can seek refuge in semantic ascent. Instead of asserting, for example, that to be Peruvian is to be from Peru, we can assert that for 'P is Peruvian' to be true is for 'P is from Peru' to be true. However, any safety that we buy in this way is illusory. Semantic ascent allows us to remain agnostic only about the form of the correct semantic theory for the claims in question; it involves no less strong a first-order metaphysical commitment.

[12] This identification in fact only holds good for the specific case of ideal gases. See e.g. M. Wilson (1985).

[13] Theoretical identifications also play key roles in other more complex explanations: for example, the identification of temperature with mean random molecular kinetic energy helps to explain why ideal gases in thermal contact will come to the same temperature.

a B, then anything we know to be an A we can also know to be a B. It is of no great importance whether we count this neutralization of why-questions by theoretical identification as exposing illegitimate demands for explanation, or as providing a positive form of explanation in its own right.

There has been plenty of debate in recent metametaphysics over what sorts of metaphysical relations 'back' reductive explanation, with a variety of notions of constitution and grounding explored in the literature.[14] My aim here is to remain neutral on this question; readers ought to be able to understand my project of reducing modality in terms of their preferred ideology. It will make no difference to the arguments of this book whether we are seeking an account of what constitutes modal facts, or of what grounds them, or of what makes them true. In places, I will use the terminology of grounding for ease of exposition but my hope is that nothing of importance hangs on this choice. Wherever a claim about grounding is used, readers should feel free to interpret it as making only a thin claim about the direction of explanation.

All this metametaphysics is no substitute for metaphysics; it is time for the main business of the book. In chapter 1, I apply the general methodology described in this section to construct a new naturalistic theory of the metaphysics of modality which draws on the resources of EQM.

[14] See, for example, Fine (2012b), J. Wilson (2014), Skow (2016), Bennett (2017), and Dorr (2016).

1
Analysing Modality

1.1 The Thesis of Quantum Modal Realism

This book is an extended defence of a metaphysical theory called *quantum modal realism*. This section sets out the main principles that constitute that theory. Here they are, in brief:

- *Diverging EQM:* Everettian quantum mechanics is correct; Everett worlds do not mereologically overlap.
- *Alignment:* To be a metaphysically possible world is to be an Everett world.
- *Indexicality-of-Actuality:* Each Everett world is actual according to its own inhabitants, and only according to its own inhabitants.
- *Propositions-as-Sets-of-Worlds:* Contingent qualitative propositions are sets of Everett worlds—a proposition P is true at an Everett world w if and only if w is a member of P.
- *Everettian Chance:* The objective chance of an outcome is the quantum weight (squared amplitude) of the set of Everett worlds in which that outcome occurs.

Why accept Diverging EQM? In chapter 2 I give some background on Everettian quantum mechanics, and there I argue for the *diverging* version of the theory according to which Everett worlds do not mereologically overlap and each macroscopic object and event exists in one Everett world only. While the correctness of EQM more generally is obviously a very substantial assumption, I shall not defend it in detail: this book presupposes EQM and explores the metaphysical implications it may have. For fuller defences of EQM from a philosophy-of-physics perspective, I instead direct the reader to Wallace (2012) and Saunders (2010a).

Why accept Alignment? I shall argue in chapter 3 that getting right the relationship between the physics of EQM and the metaphysics of modality is crucial to an adequate Everettian treatment of objective probability. A single theoretical identification is needed to close the circle and connect

The Nature of Contingency: Quantum Physics as Modal Realism. Alastair Wilson, Oxford University Press (2020).

DOI: 10.1093/oso/9780198846215.001.0001

up weight and chance in the right way—the identification of alternative metaphysical possibilities with Everett worlds. Once this identification is made, the way lies open to a rich and powerful new theory of the nature of modality, one which provides significant further reasons for Everettians to adopt Alignment.

What is an Everett world? It is a global quantum-mechanical sequence of events, and the key theoretical device of modern EQM. An immediate consequence of EQM is that we know Everett worlds by direct acquaintance: the concrete physical universe around us is an Everett world. EQM—characteristically, strikingly—posits more universes of the same kind as ours. The Everett worlds, taken together, comprise the Everettian multiverse. No two Everett worlds are qualitatively indiscernible: indeterministic quantum events turn out differently in different worlds, and there is a world for every quantum-mechanically possible sequence of events.

As I understand Everett worlds, they are causally isolated. Causal relations obtain only between events within individual worlds, and there are no causal relations between events in different worlds; travel between Everett worlds, sadly, remains in principle impossible. A real physical quantity, *amplitude*, provides a chance measure defined over the whole space of Everett worlds.[1]

A more detailed description of the physics and metaphysics of Everett worlds is given in sections 2.2–2.4. An intuitive picture of these worlds will suffice for the arguments of this chapter. I recommend that for the time being readers think of Everett worlds as parallel universes that fit together into one giant jigsaw, filling up possibility space with all of the different histories that physics permits, with each world a complete four-dimensional spacetime that is governed—from the perspective of its inhabitants—by indeterministic laws. One of these four-dimensional spacetimes is our own, and the others are things of the same general kind.

Before we can properly assess Alignment, we need also to say more about what is meant by 'metaphysically possible'. I take it that recent metaphysics has characterized a fairly distinctive theoretical role for metaphysical modality: at its core it is intended to be an objective modality tied to a distinctive class of counterfactuals (Lewis 1986b, Williamson 2007) and to the necessary a posteriori (Kripke 1980), and with connections to other notions like

[1] Formally, amplitude is a complex number and the quantity identified with probability via the Born rule is the modulus-squared amplitude. For simplicity, I will use the terms 'amplitude' or 'weight' interchangeably to refer to the measure over histories given by the Born rule.

conceivability and essence which are perhaps more negotiable. What best ties these features together, I suggest, is that they are features that we would expect of *whichever objective modality turns out to be metaphysically most perspicuous*. This descriptive way of fixing the reference of 'metaphysical modality' has some substantive realist presuppositions—deflationists about modality such as Sider (2011) and Cameron (forthcoming) will likely deny that there is any such thing as the most metaphysically perspicuous objective modality. Still, such a modality is my analytical target in this book; I shall argue that the core roles of metaphysical modality can be accounted for using the theoretical resources of Everettian quantum mechanics.

Taking metaphysical modality as my analytical target will require some deviations here and there from mainstream opinions about metaphysical modality, but that is inevitable in an analysis which is intended as theoretically revisionary. What matters is that enough of the core features of metaphysical modality are captured. For those who remain unconvinced that the modality I shall characterize in quantum terms deserves the term 'metaphysical possibility', I offer a different way of looking at the matter. We may regard quantum modal realism as eliminating the category of metaphysical modality and instead employing a combination of physical modality and logical modality to recapture some of the explanatory work previously assigned to metaphysical modality. The differences between these two approaches are largely terminological and I will not dwell on them any further: 'metaphysical possibility' is after all a term of art, and I am primarily interested in the substantive connections between quantum physics and chances, counterfactuals, causation and the like.

Alignment entails the following two principles:

Individualism: If X is an Everett world, then X is a metaphysically possible world.
Generality: If X is a metaphysically possible world, then X is an Everett world.

Individualism is one of a number of options we have when choosing how to think about the connection between metaphysical modality and EQM.[2] The

[2] Individualism is very much a minority view, whether in the context of EQM or in the context of multiverse theories more generally; as I shall argue in chapter 3, much work on the metaphysics of EQM and of other multiverse theories uncritically presupposes Collectivism. Nonetheless, at least one established philosophical tradition seems to invoke the combination of Individualism with the denial of Generality. This is the branching time (or, in its relativistic form, branching spacetime) programme, whose genesis owed much to Saul Kripke and to

most natural alternative would be to take Everettian multiverses, rather than Everett worlds, to play the theoretical role of alternative metaphysically possible worlds in the Everettian paradigm:

Collectivism: If X is an Everettian multiverse, then X is a metaphysically possible world.

Individualism and Collectivism say very different things about the structure of modal reality. According to Collectivism, Everettian multiverses are metaphysically possible worlds which have the peculiar property of having many parts, each of which is isomorphic to a possible world of the sort envisaged by the classical one-world conception of the relation between physical theory and metaphysical modality. According to Individualism, Everettian multiverses are complexes of metaphysically possible worlds; they are not themselves identical to or located within any single metaphysically possible world.

Why accept Individualism? I will argue in chapter 3 that it does crucial work when it comes to giving a positive theory of Everettian probability. However, there is a presumption (mostly implicit) in favour of Collectivism over Individualism amongst those who have thought and written about EQM. I suggest that this state of affairs can be traced back to the following a priori assumption:

Physical Actualism: Each model of a complete physical theory represents exactly one possible world.[3]

Physical Actualism—combined with the relatively uncontroversial consequence of decoherence theory that models of EQM have the structure of a multiverse—straightforwardly entails Collectivism. But Physical Actualism

Arthur Prior (the idea was mooted in Kripke's letter to Prior of 3 September 1958) but which is now strongly associated with Nuel Belnap and his followers (see e.g. Belnap et al. 2001) and which has been applied to EQM by Belnap & Muller (2010). Fans of branching (space-)time typically want to allow deterministic worlds as a special case—as a very simple branching structure which happens to have no nodes—but they allow that indeterministic worlds are represented by individual paths through a branching network, rather than by the entire network.

[3] In the presence of redundancy in our physical description, multiple models of a theory may correspond to one single possible world. This is beside the point for present purposes—what matters is whether one single model of the physical theory may correspond to multiple possible worlds.

is much less natural in the context of 'many-history' physical theories, such as EQM, than it is in the context of 'single-history' theories, such as classical mechanics. And, as I shall argue in chapter 3, it prevents us from accommodating non-trivial objective probability in the Everettian picture. I take this to be sufficient reason to reject Physical Actualism in full generality, clearing the way for us to accept Individualism and the resulting elegant treatment of quantum chance.

Why accept Generality? I will argue for this principle by appeal to the theoretical unity and simplicity of the systematic metaphysics that it makes possible. Without Generality, Everettians must distinguish two fundamental and fundamentally different kinds of possibility; Generality provides theoretical uniformity. Generality also enables a wholly reductive theory of objective modality, and a straightforward account of modal epistemology which renders it continuous with ordinary scientific inquiry. These results serve to underwrite applications of objective modality in the metaphysics of laws, properties and causation, and in the semantics of counterfactuals. In sum, a theory of modality incorporating Alignment—by entailing both Individualism and Generality—has decisive theoretical advantages over a theory of modality incorporating Individualism without Generality.

Adopting Alignment amounts to equating physical possibility (necessity) with metaphysical possibility (necessity)—or at least taking them to be coextensive. This equation has recently grown in popularity, with various versions of *necessitarianism* about laws being defended by (inter alia) Shoemaker (1980, 1998), Swoyer (1982), Fales (1993), Ellis (2001), Bird (2001, 2004, 2007), and Edgington (2004). The version of necessitarianism on which I will focus is what Schaffer (2005) calls *modal necessitarianism*: the view that the laws of the actual world are the laws of all possible worlds. For arguments in support of modal necessitarianism which are largely independent of EQM, see A. Wilson (2013b). However, I think the overall explanatory power of quantum modal realism provides the most compelling motivation so far adduced for modal necessitarianism, or indeed for any form of necessitarianism. Quantum modal realism also provides unique ways to sweeten the pill of accepting modal necessitarianism, as I argue in section 4.6.

Why accept Indexicality-of-Actuality? The principle could be defended on grounds of ideological simplicity since it allows us to dispense with a notion of absolute actuality, or on grounds of avoiding sceptical worries since it neutralizes the concern that we ourselves might not be actual. But my main argument for the principle is that it simply falls out of the combination of Alignment with Diverging EQM: I include it as a separate principle only for clarity. It is already a key part of the conceptual basis of EQM that all Everett worlds are on a par, none physically or ontologically privileged in any way; Everett (1957b: 3) was explicit about this aspect of the theory. And if the Everett worlds are the metaphysically possible worlds, then it immediately follows that all metaphysically possible worlds are on a par, none physically or ontologically privileged in any way. Indexical actuality is the only notion of actuality that properly respects this constraint. The consequences of quantum modal realism for the concept of actuality are explored further in section 1.9.

Why accept Everettian Chance? Here I simply refer the reader to chapter 3, which contains an extensive discussion of the theoretical role of objective chance and the extent to which the quantum weights can play that role. To foreshadow, the argument is that the quantum weights are the clear best candidates to play the chance role within the context of EQM—so, following the metaphysical methodology set out in the introduction, weights *are* chances.

David Lewis (Lewis 1986b) offered a dizzying array of philosophical explanations based on the resources of his own reductive analysis of metaphysical modality. In addition to their central application to the analysis of modal thought and talk (Lewis 1986b section 1.2), modal realist resources are applied to the semantics of counterfactuals and hence to the semantics of causal and dispositional claims (ibid. section 1.3), to the analysis of semantic and mental content in general (ibid. section 1.4), and to the metaphysics of properties and laws (ibid. section 1.5). Despite the differences between quantum modal realism and Lewisian modal realism, we can capture the key achievements of the Lewisian approach in the quantum modal realist picture. In the remainder of this chapter, I give an argument for quantum modal realism from overall theoretical virtue that echoes the structure of Lewis's argument for Lewisian modal realism.

While not all of the Lewisian analyses of modal notions carry over directly to the Everettian context, quantum modal realists can in each case adapt the core idea. The resulting quantum modal realist analyses retain the most compelling feature of the Lewisian analyses: their genuinely reductive character. As I read him, most central to Lewis's case for modal realism was his dissatisfaction with primitive modality. Motivated by a principle of dependence of truth on being, Lewis wanted all truths to be accounted for in terms of how things are. Modal realism is an implementation of this broader project, by explaining modal truths in terms of non-modal truths about the configuration of objects across modal space. The theory explains how things possibly are and necessarily are in terms of how things are *simpliciter*: we do not need to expand our primitive ideology in order to understand modality. Quantum modal realism likewise enables an explanatory account of modal truth in terms of non-modal being.

Although Lewis rather puzzlingly claimed to regard the lack of arbitrariness in the modal realist worldview as a problem (Lewis 1986b: 128), he made regular appeal to avoidance of arbitrariness elsewhere in his theorizing; and in any case there is a long tradition, tracing back at least to Parmenides, of viewing avoidance of arbitrariness as an important virtue in fundamental metaphysics. Lewisian modal realism scores spectacularly well on this count: it seems to render reality as a whole entirely non-arbitrary. The principle of recombination applies uniformly with no exceptions, no arbitrary 'gaps in logical space'. Quantum modal realism likewise minimizes arbitrariness: the Schrödinger equation applies universally, with an Everett world for every quantum-mechanically possible history. There are no arbitrary gaps in the multiverse. All arbitrariness in both forms of modal realism is reduced to perspectival arbitrariness: why did I occupy this particular perspective rather than any other?

One place in which arbitrariness might seem to remain within quantum modal realism is in the initial quantum state of the universe. Since quantum modal realists model contingency as variation across Everett worlds, there can be no contingency in an initial state that these worlds have in common. It remains an open theoretical and empirical question, however, what this initial quantum state is like: different candidate frameworks for a fundamental theory, such as string theory or loop quantum gravity, are characterized by very different-looking initial states. If it were to turn out that the true initial quantum state of our universe has arbitrary-seeming features that lack any apparent theoretical explanation, that would be prima facie evidence against quantum modal realism—since it would suggest a source of contingency in reality that goes beyond quantum contingency. But at present there is no reason to believe this is how things will turn out. (See chapter 6 for more discussion of potential contingency in overall cosmology.)

1.2 Quantum Modal Realism at Work: Modality

The basic form of the quantum modal realist analysis of modality is encoded in the principles Alignment and Indexicality-of-Actuality. Everett worlds are metaphysically possible worlds, and our own Everett world is the actual world. So, at a first pass: for an event to be metaphysically possible is for it to occur in some Everett world, for it to be metaphysically necessary is for it to occur in all Everett worlds, and for it to be actual is for it to occur in our own world. This account of modality will be refined and reformulated over the course of this section, but we can already make out some of its general features. In particular, the space of Everett worlds is characterized in quantum-theoretic terms independently of the notion of metaphysical modality, and consequently quantum modal realism is a reductive theory: it accounts for metaphysical modality without presupposing any modal notions.

There are fewer worlds encompassed in quantum modal realism than there are in the more familiar Lewisian modal realism. No worlds that conflict with quantum physics exist. Still, we have Everett worlds for all the most obviously possible courses of events. To enumerate exactly which Everett worlds there are and what they are like, we would need a fully completed science; but even with existing scientific knowledge we can have plenty of substantive knowledge about the space of Everett worlds. The roles of intertheoretic relations and of special science knowledge are key. For example, consider the possibility of kangaroos without tails. We know what sorts of macroscopic configurations of materials a kangaroo corresponds to, we know what sorts of microscopic configurations of atoms and molecules are capable of realizing those macro-configurations, we know that quantum theory permits microscopic configurations of that sort, and we know that physical and chemical processes that would generate such configurations are assigned non-zero quantum-mechanical chances. In contrast, we know of no physical processes that could give rise to ghosts of people from the distant past (as opposed to hallucinations of ghosts, or to ghost-shaped clouds of dust), so we have no reason to think that there are any ghosts in any Everett world.

The quantum modal realist, like the Lewisian modal realist, may use restrictions on the space of metaphysical possibilities to characterize interesting *restricted modalities* that can serve various theoretical purposes. From the base space of Everett worlds—the entire plurality—we can extract any sub-plurality and quantify only over those worlds to generate a new restricted modality. For example, restricting to worlds with initial segments that match the actual past generates a notion of historical modality: it is historically possible that humanity should continue to flourish for a million years, but not historically possible that World War II should not occur. In possible-worlds semantics for modal logic,[4] varying the *accessibility relation* implements such restrictions and allows clean characterization of any number of distinct modalities. Some of the more interesting restricted modalities will make an appearance in chapter 4.

As noted in section 1.1, quantum modal realism collapses two types of modality that are usually regarded as distinct: physical modality and metaphysical modality. Consequently, quantum modal realists have no need of the Lewisian account of physical modality as a restricted modality with the accessibility relation fixed by the laws of nature. However, it is no deficiency of quantum modal realism that it does not treat physical modality as a restricted modality. All realists about objective modality must accept some space of worlds as the most inclusive, and decline to give a treatment of this basic modality as a restriction on any further modality. Moreover, realist theories can still consistently incorporate more inclusive forms of possibility, as long as these more inclusive modalities can be given some non-realist interpretation. Lewisian modal realism is not itself inimical to a broader notion of logical possibility where the only necessary truths are logical truths, although the Lewisian modal realist does owe us some analysis of this logical possibility in terms of metaphysical possibility. Similarly, quantum modal realists can offer constructions of various logical and conceptual modalities. These matters are considered further in section 1.4.

As Lewis remarked, 'modality is not all diamonds and boxes'. The claims of *de dicto* possibility, necessity, and actuality that we have so far considered are only a small fragment of our modal thought and talk. In section 1.3 of Lewis (1986b), Lewis mentions a number of modal idioms which are hard to capture using familiar quantified modal logic, but which are straightforward to capture by quantifying in creative ways over the individuals in some range of Lewis-worlds. Quantum modal realists can likewise quantify over a limited range of individuals in a limited range of Everett worlds, to make sense of claims such as 'there are three possible ways for this plan to succeed' and 'a mammal could resemble a bird more closely than a mammal could resemble an amoeba'. The results of this quantification in the Everettian theory will of course not align exactly with the results in the Lewisian theory, since the space of worlds quantified over differs in extent; but the structural features of the analysis are the same, making it equally flexible.

[4] For introductions to possible-worlds semantics, see Divers (2002) and Melia (2003).

A key notion which Lewis analyses directly by way of his modal realism is *supervenience*. Recent metaphysics has found much use for this notion, and numerous different flavours of supervenience have been discussed; McLaughlin and Bennett (2018) give a comprehensive summary. The question of whether the mental facts supervene on the physical facts has long been a popular way to draw the battle-lines between dualists and materialists; Humeans and non-Humeans in the metaphysics of laws of nature have argued over nomological supervenience theses such as Lewis's *Humean Supervenience*; and metaphysically inclined philosophers of language dispute whether semantic facts supervene on natural facts.

The account of supervenience that I will offer is, in effect, identical to Lewis's. Again, it is given directly in terms of the reductive base rather than via a detour through a formal language containing modal operators. Lewis (1986b) argued that analyses of supervenience that make use of modal logic's familiar diamonds and boxes can express what he calls 'narrow' supervenience but not 'broad' supervenience:

> 1) *Narrow* psychophysical supervenience: could two people differ mentally without also themselves differing physically?
>
> 2) *Broad* psychophysical supervenience: could two people differ mentally without there being a physical difference somewhere, whether in the people themselves or somewhere in their surroundings?
>
> Lewis (1986b: 14–15)

Both of these notions are easy enough to express if we allow ourselves the modal realist's quantification over possible worlds:

> Among all the worlds, or among all the things in all the worlds, . . . there is no difference of the one sort without difference of the other sort. Whether the things that differ are part of the same world is neither here nor there.
>
> Lewis (1986b: 17)

According to this account of supervenience, narrow psychophysical supervenience is the thesis that no world contains two things which differ mentally without differing physically; and broad psychophysical supervenience is the thesis that no two worlds differ in any of the mental properties instantiated at them without differing in the physical properties instantiated at them.

Narrow psychophysical supervenience can be modelled using a modal sentential operator: it is the thesis that it is not the case that, possibly, there are two things which differ mentally without also differing in their intrinsic physical character. Lewis argues, however, that the closest we can get to broad psychophysical supervenience using a sentential operator is the following: it is not the case that, possibly, there are two things which differ mentally without differing in their intrinsic or extrinsic physical character. But then:

> What we have said is not quite what we meant to say, but rather this: there could be no mental differences without some physical difference *of the sort that could arise between people in the same world*. The italicised part is a gratuitous addition. Lewis (1986b: 16)

He counts it a significant advantage of his modal realism that it allows both notions to be stated, and to be distinguished from one another: 'The moral is that we'd better have other-worldly things to quantify over—not just a primitive modal modifier of sentences' (Lewis 1986b: 17). Quantum modal realism provides us with the desired other-worldly things to quantify over; so it is no surprise that the Lewisian analysis of supervenience carries over unmodified to quantum modal realism. An Everettian broad psychophysical supervenience thesis says that no two Everett worlds differ in any of the mental properties instantiated at them without differing in the physical properties instantiated at them.

One central application of the notion of supervenience is to characterize global metaphysical outlooks. The most familiar such thesis is David Lewis's doctrine of Humean Supervenience:

> All there is to the world is a vast mosaic of local matters of particular fact, just one little thing and then another . . . We have geometry: a system of external relations of spatiotemporal distances between points . . . And at those points we have local qualities: perfectly natural intrinsic properties which need nothing bigger than a point at which to be instantiated. For short: we have an arrangement of qualities. And that is all. There is no difference without difference in the arrangement of qualities. All else supervenes on that. Lewis (1986a: ix)

Humean Supervenience is intended to rule out ingredients of reality—disembodied minds, causally epiphenomenal ectoplasm, and the like—that float free from the sort of field-theoretic ontology which he took to be suggested by fundamental physics. Humean Supervenience incorporates a thesis about the fundamental elements of reality—they are intrinsic properties instantiated at spacetime points—but the philosophically boldest part of the thesis is the claim that all phenomena of any kind at any scale asymmetrically modally depend on phenomena of the kind described by fundamental physics. Despite Lewis's regular association with the worst excesses of analytic metaphysics, Humean Supervenience was proposed in a spirit of respect for the ontology of physics and for the foundational role of physics in constituting and constraining special-science phenomena.

Humean Supervenience is not the only global supervenience thesis in town. Alternatives include Bigelow's liberal thesis that 'truth supervenes on being' (Bigelow 1988: 132–3), Armstrong's Aristotelian-style supervenience thesis that all facts supervene on facts about which entities instantiate which universals (Armstrong 2010), Chalmers and Jackson's A Priori Entailment Thesis (Chalmers and Jackson 2001), and Chalmers's more recent variety of 'scrutability theses' (Chalmers 2012). We need not delve into the details: all of these theses quantify over metaphysically possible worlds, and so all of them carry over straightforwardly to the Everettian context. Alignment tells us that Everett worlds are the metaphysically possible worlds, and accordingly Humean Supervenience and its kin *just are* theses about variation across the space of Everett worlds. Which of the various supervenience theses should quantum modal realists endorse? I leave this question open here.

There are plenty of other philosophically interesting notions in the vicinity of supervenience. Instead of asking—say—what the laws of nature supervene on, we can ask what determines the laws, or what fixes them, or what grounds them, or what they depend on, or in virtue of what they are the laws, or what makes them the laws. Which of these questions you take to be most interesting will depend on your preferred metametaphysical ideology. Making sense of many of the questions requires recognizing 'worldly' (rather than merely representational) hyperintensional distinctions—distinct yet necessarily equivalent facts or properties—which goes beyond the objective modal distinctions of either Lewisian modal realism or quantum modal realism. However, defenders of strong notions of grounding or ontological dependence typically regard them as primitive; quantum modal realists can do likewise if they wish, and are accordingly no worse off than anyone else with respect to making sense of generalized determination relations. Further discussion of these matters will be deferred to section 1.4.

Lewis offered a theory of coarse-grained propositions as sets of worlds. For the purposes of resolving the probability problem in EQM, chance-bearing propositions will in chapter 3 be similarly analysed in terms of sets of Everett worlds. This analysis is fine for ordinary factual propositions of the kind we use most frequently to communicate in everyday life and scientific practice. However, we may legitimately worry that the analysis will be inapplicable to the propositions expressing various parts of EQM, such as the Schrödinger equation itself. The structure of this difficulty is not, of course, unique to EQM; it has been discussed in the context of Lewisian modal realism under the rubric of *advanced modalizing*. Quantum modal realism is a form of modal realism—it says that multiple alternative possibilities genuinely exist—so it is unsurprising that it faces an analogous problem with advanced modalizing.

In the remainder of this section, I will explain the challenge of advanced modalizing, and discuss some potential responses to it—initially in the context of Lewisian modal realism. My preferred account of advanced modalizing will afterwards be adapted to the setting of quantum modal realism. The resulting conception of the modal status of the fundamental facts vindicates the necessity-first approach to analysing modality that I endorsed in section 0.3 of the introduction.

Advanced modal claims, as characterized by Divers (1999) in the context of Lewisian modal realism, are modal claims about entities that are spatio-temporally unrelated to us. For most metaphysicians, this will include modal claims about entities sometimes described as 'necessary existents', such as numbers, sets, properties, and so on. But they will also include modal claims about individuals in other Lewisian worlds. A Lewisian modal realist is committed to the existence of such individuals; others might be committed to their non-existence. But whether we believe in such individuals or not, we can still ask about their modal status; we can ask, in particular, whether their existence is a contingent matter or a necessary matter. If there is a Lewisian pluriverse, could there have failed to be one? And if (as most of us believe) there is no Lewisian pluriverse, could there have been one?

Lewis's original statement of Lewisian modal realism was given in 'Counterpart theory and quantified modal logic' (Lewis 1968). The view was there presented as a way of providing truth-conditions for expressions in a formal language containing modal operators (the universe language) using expressions in a formal language lacking modal operators (the pluriverse language). This original, explicitly metalinguistic, characterization of modal realism has perhaps helped to obscure the status of advanced modalizing.

If modal realism were a purely metalinguistic doctrine, then it would be open to Lewis to simply deny the coherence of asking advanced modal questions. Sentences which combine elements of the universe language (modal operators) with elements of the pluriverse language (quantification over entities in other worlds) would themselves be sentences of no well-defined language (or at least, no metaphysically perspicuous language), and could be safely neglected. Thus the metalinguistic formulation of modal realism leaves no room even to pose the questions involved in advanced modalizing. Questions like 'could what unrestrictedly exists have been different?' can then be ignored as meaningless (or as metaphysically unperspicuous), since they illegitimately mix the universe language and the pluriverse language.

The metalinguistic escape route cannot, in the end, be sustained. Semantic ascent provides only temporary respite from the commitments of first-order discourse: when we try to explain the usefulness of that discourse, or to explain what grounds the truth of some metalinguistic theory, we will be forced to revisit the metaphysical questions that semantic ascent promised to bypass. Lewis acknowledged this methodological point, and his later presentations of modal realism—in *Counterfactuals* (Lewis 1973) and in *On the Plurality of Worlds* (Lewis 1986b)—are formulated in plain English (or at least in the philosopher's semi-formalized dialect of English) without any metalinguistic garb.

Once Lewis abandoned the metalinguistic 'two languages' formulation of modal realism, he required a different mechanism to distinguish the metaphysician's discourse about the whole plurality of worlds from more ordinary discourse about goings-on within some individual world. The mechanism he adopted was *quantifier restriction*. 'When they attach to sentences we can take the diamond and the box as quantifiers, often restricted, over possible worlds' (Lewis 1986b: 9). Lewis's presentation in *On the Plurality of Worlds* is deliberately informal, with no emphasis on the logic of modal operators. Indeed, Lewis goes out of his way to make some scornful remarks about 'boxes and diamonds' (ibid.: 13). This feature of the presentation has, I think, disguised a significant ambiguity in the way that quantifier restriction is supposed to work for the modal realist. It is usually thought that Lewis's analysis of modal operators works by restricting quantifiers to range over entities in individual worlds, as intimated by the 'at W' locution. Lewis invited this understanding of his view with informal statements of his analyses such as the following: '[P]ossibly there are blue swans iff, for some world W, at W there are blue swans' (Lewis 1986b: 5). The example seems unproblematic: the expression 'at W' restricts quantifiers to range only over entities which are parts of W. So if it is (unrestrictedly) true that a world W contains blue swans, then it will be true that *at W there are blue swans.*

The problem is that analyses like these threaten the well-understood logical behaviour of the modal operators. For example, the orthodox T axiom ($\Box P \rightarrow P$) and the duality of the box and diamond together entail $P \rightarrow \Diamond P$. Now let P be the statement that there exists the Lewisian plurality. Modal realists, of course, accept P. So if they also accept axiom T, then they must accept that it is possible for the Lewisian plurality to exist. But if we feed this claim into the Lewisian analysis of 'possibly φ' quoted above, then we get the following problematic consequence:

Worlds-within-worlds: Possibly there exists a Lewisian plurality iff, for some world W, at W there exists a Lewisian plurality.

If, as previously suggested, 'at W' restricts quantifiers to range over only entities which are part of W, then Worlds-within-worlds comes out false. There is no world that contains a plurality of worlds. But then it looks like the modal realist analysis has delivered the wrong result, by entailing that a Lewisian plurality is impossible.

This problem has recently begun to attract more attention (see Divers 1999, Parsons MS, Dorr MS, Marshall 2016, Jago 2014). Josh Parsons argues that the best response for the modal realist is to modify the analysis of 'possibly φ' so as to render it disjunctive. He calls this the 'T-preserving analysis', and states it as follows (S is an arbitrary sentence):

> $\ulcorner S \urcorner$ is possibly-true iff S, or for some world w, $[S]^{w}$.
>
> $\ulcorner S \urcorner$ is necessarily-true iff S, and for every world w, $[S]^{w}$.
>
> As far as the analysis of possibility and necessity goes, this is equivalent to taking the whole pluriverse to be an extra, large world of many island universes. Parsons (MS: 8)

This obtains the right result in our test case—it comes out true at any world that there could be the Lewisian plurality—but it is an unattractive response to the problem of advanced modalizing. For a start, it looks ad hoc—there is no motivation for the added disjunct other than to solve the problem in question. Disjunctive analyses are commonly thought to be less satisfactory than non-disjunctive analyses, *ceteris paribus*.[5] The real problem with Parsons's proposal, though, is that it is insufficiently general. It allows that it is possible for there to be a Lewis-world, and for there to be a Lewisian plurality, but not for there to be any other number of Lewis-worlds. It will be necessarily true that either there exists the entire Lewisian plurality, or that there exists exactly one Lewis-world. This consequence is unpalatable enough to reject the T-preserving analysis if any better alternative can be found.

The problem with the T-preserving analysis could be resolved by adopting an idea from Bricker (2001) and allowing arbitrary mereological sums of Lewis-worlds to count as Lewis-worlds. Then it would be possible for there to be a Lewisian plurality, possible for there to be a single Lewis-world, and possible for there to be any number in between. (It would however be impossible for there to be no Lewis-worlds at all.)

So, to summarize, we find disagreement over what a modal realist should say about the modal status of the claim that the Lewisian plurality exists. Lewis appears to accept that it is necessarily true. Parsons argues that the modal realist should say that it is contingently true, and adopting Bricker's proposal has the same consequence. Perhaps most counter-intuitively of all, Noonan (1994) and Hudson (1997) argue that modal realism, if true, is necessarily false.

What I think the modal realist ideally needs is an elegant way of vindicating the natural view of the plurality as non-contingent. Not only does the thought that the Lewisian plurality necessarily exists fit nicely with the thoroughly orthodox view that metaphysical truths are necessary truths, but rejecting it will inevitably involve rejecting the strong and attractive S5 modal logic for metaphysical modality. Fortunately, there is a way of underwriting the Lewisian view of the modal status of the claim that the Lewisian plurality exists, which can respect axiom T and retain S5 in full generality. This option is defended as the best version of modal realism by Dorr (MS) and by Marshall (2016); certain features of it were anticipated by Divers (1999). It involves generalizing the idea that *de re* modality is explicable in terms of counterpart relations to the idea that all modality, including *de dicto* modality, is so explicable.

On my favoured counterpart-theoretic semantics for modal realism, the modal operators 'possibly' and 'necessarily' are understood as quantifiers over

[5] Why this is the case is a very interesting question, which is addressed in Dorr (2010).

generalized counterpart relations—or, as Dorr calls them, *counterpairings*. Counterpairings are functions from the individuals in the pluriverse to the individuals in the pluriverse: they include intricate permutations of all individuals, permutations of just two individuals that leave all others unchanged, and the identity function. 'Possibly, *a* is *F*' says that there exists a counterpairing that maps *a* to something that is *F*. 'Necessarily, *a* is *F*' says that all counterpairings map *a* to something that is *F*. As in classical possible-worlds semantics, the duality of the existential and universal quantifiers guarantees the duality of the possibility and necessity operators. And, as Dorr (MS) shows, some straightforward restrictions on which counterpairings are admissible—in particular, the requirement that counterpairings be *global permutations* of the whole domain—are enough to ensure that all the features of S5 modal logic are recovered.

To extend this counterpart-theoretic treatment from *de re* modality to encompass *de dicto* modality, we may treat ordinary claims in ordinary contexts as attributing some property to the actual world; modalizing such claims then requires counterpairings that permute whole worlds. 'Necessarily, p', by the counterpart-theoretic semantics, becomes 'all counterpairings map the actual world to a world such that, restricting our quantifiers to that world, p'. Likewise, 'possibly p' is rendered as 'some counterpairing maps the actual world to a world such that, restricting our quantifiers to that world, p'.

Discourse about the pluriverse itself is not ordinary discourse, and claims about the pluriverse are not attributions of any property to the actual world. Modalizing claims about the pluriverse, in the counterpart-theoretic approach, instead requires counterpairings that permute whole pluriverses. However, since there is just one pluriverse, there is just one way of mapping pluriverses to pluriverses: so all counterpairings map the pluriverse to itself and not to anything else. 'Necessarily, there is a pluriverse' is then equivalent to 'there is a pluriverse', which in turn is equivalent to 'possibly, there is a pluriverse': modal operators are redundant when applied to quantifiers that are already completely unrestricted. I shall call this approach the *counterpairing theory* of advanced modalizing (to distinguish it from the simpler 'redundancy theory' of Divers (1999) criticized by Parsons, Marshall, Jago and others). Applied to the claim that there (unrestrictedly speaking) exists a pluriverse, the counterpairing theory delivers the result that the existence of the pluriverse is both possible and necessary.

There is good reason to think that Lewis endorsed something like the counterpairing theory. He replies to an objection based on the knowability of other Lewis worlds by declaring that our knowledge of the other worlds is *necessary knowledge*, and as such does not require causal contact with them; this certainly suggests that he thinks of the plurality as existing necessarily. And he says other things which suggest that modal reality is non-contingent. For example, he allows that the following are consequences of modal realism:

> There is but one totality of worlds; it is not a world; it could not have been different. Lewis (1986b: 80)

> all statements about modal reality [are] non-contingent: if any such statement is possibly true, it is true *simpliciter*. Lewis (1996: 683)

This certainly sounds like the counterpairing theory. Dorr (MS) assembles some further evidence that Lewis endorsed this approach to advanced modalizing.

In the light of the simplicity and the attractive logical properties of the counterpairing theory, it is charitable to interpret Lewis's statement that 'possibly there are blue swans iff, for some world *W*, at *W* there are blue swans' (Lewis 1986b: 5) as merely a simplified first-pass analysis. This is after all Lewis's first statement of the modal realist analysis of modality in *On the Plurality of Worlds*, and he is more careful in later passages. Once we move from this rough first-pass analysis to the counterpairing view based on generalized counterpart relations, then the path is cleared for a modal realist (of either the Lewisian or quantum variety) to theorize freely about the plurality of worlds while retaining all of the standard logical machinery of S5.

The discussion in this section has been of advanced modalizing in Lewisian modal realism, but nothing crucial has turned on the distinctive features of the Lewisian ontology. Any version of modal realism—any account according to which alternative possibilities exist and are of a kind with the actual world—can avail itself of the counterpairing theory to account for the modal status of the ontological claims which it itself incorporates. In particular, quantum modal realism can fruitfully be combined with the counterpairing theory of advanced modalizing.

To summarize: quantum modal realists require an account of advanced modalizing in order to coherently state their theory. Various such accounts are available, and debate continues over which of them are tenable; quantum modal realists could in principle adopt any viable such account. However, the counterpairing theory recommended by Dorr to Lewisian modal realists has some powerful advantages, especially with respect to vindicating the connection between the modal realist metaphysics of modality and S5 modal logic. Quantum modal realists may adopt this theory, and by doing so they avoid the most prominent potential worries about the coherence of their framework.

1.3 Quantum Modal Realism at Work: Closeness

One of the most prominent roles that a theory of modality needs to play is that of providing truth-conditions for a semantic analysis of counterfactuals. In this section, I argue that quantum modal realism provides extremely rich resources for such an account—richer in certain respects than the resources of Lewisian modal realism—and that quantum modal realism also allows for a uniquely satisfying explanation of why counterfactuals should be indispensable in our practices of giving explanations and making predictions.

According to quantum modal realism, there just are no deterministic possible worlds (chapter 4 provides more details); so there is no need for a theory of counterfactuals to deliver plausible truth-conditions even under determinism. This observation paves the way for semantic theories of counterfactuals that rely essentially on indeterminism. I shall describe two alternative approaches to building such semantic theories: a primary approach which adapts the closest-worlds approach taken by Lewis, and a secondary approach which bypasses closeness orderings altogether to theorize directly in terms of (contextual) conditional chances.

There is one core idea driving the hugely influential Stalnaker–Lewis possible-worlds semantics for counterfactuals (Stalnaker 1968; Lewis 1973): it is the idea that counterfactual conditional assertions are statements *about how things stand with respect to other genuine possibilities.* When I say 'if kangaroos had no tails, they would fall over', the Stalnaker–Lewis account represents me as talking directly about the tribulations of some possible kangaroos (Lewis 1973: 1). Quantum modal realists can and should incorporate this same core idea. Since quantum modal realists identify Everett worlds with genuine metaphysically possible worlds, the quantum modal realist version of the Stalnaker–Lewis approach has it that counterfactual conditional assertions tell us about how things stand with respect to Everett worlds other than our own. When we engage in counterfactual discourse, what we are doing is charting variation in occurrent fact across the quantum multiverse.

Much of the controversy around the Stalnaker–Lewis semantics is the result of its reliance on facts about other possible worlds, construed as conceptually prior to our counterfactual thought. Since almost nobody currently endorses Lewisian modal realism, mainstream versions of the Stalnaker–Lewis semantics interpret counterfactual conditionals as assertions about how things stand with respect to some ersatz world: to a set of propositions, say, or to some entity whose nature is to represent primitively. (See section 1.3 for further discussion of ersatzist views.) This insubstantial metaphysical foundation is a source of ongoing dissatisfaction with contemporary understandings of the Stalnaker–Lewis semantics: see, for example, Kripke (1980) and J. Wilson (2014). Even if this metaphysical queasiness can be quelled, the ersatzist versions of the Stalnaker–Lewis semantics sacrifice much of the plausibility of the core idea driving the semantics, that counterfactual discourse is about how things stand with respect to genuine alternative possibilities. We undeniably talk often and care deeply about what would and could have been; but there is no plausible story to be told about why we should talk so often and care so deeply about how some false set of propositions represents things as being.

Can the same objection be made to quantum modal realism? Does learning that possible worlds are Everett worlds undermine the plausibility of the view that counterfactual thought is thought about goings-on within them? The answer is 'no'. Everett worlds, on the present proposal, are things with one of which we are already intimately acquainted: the actual world is an Everett world, and the other Everett worlds are simply more things of exactly the same basic kind. Accordingly, in the context of quantum modal realism the core Stalnaker–Lewis idea that counterfactual thought is about goings-on in other genuinely possible worlds is implemented in such a way that counterfactual thought is thought about goings-on within things that are fundamentally of the same kind as the actual world. (It might better be thought of as *counteractual* thought, since there are real (non-actual) facts about other Everett worlds.)

Providing entities that are good candidates to be the subject matter of counterfactual discourse is only the first step towards a full theory of counterfactuals. The Stalnaker–Lewis account by itself does not give us any direct guidance as to which counterfactuals are true, or even uniquely pin down a precise logic for counterfactuals. In order to proceed further, we need to choose between the Lewisian implementation of the Stalnaker–Lewis semantics (with a context-dependent *closeness ordering* picking out some set of closest worlds) and the Stalnakerian implementation of that semantics (with, in addition, a context-dependent *selection function* picking out a single closest world). I follow the closeness-ordering approach here, and show how Lewis's account of closeness needs to be adapted to the quantum modal realist setting.

A central motivation behind the Lewisian account of the closeness ordering was to account for counterfactual truth in deterministic worlds without radical *backtracking*. Even the simplest counterfactual antecedent, if it is to be implemented compatibly with deterministic laws, requires tracking all the way back to the initial condition of the universe. Given determinism, the only physically possible way for things to be different now is for them to have been different all the way back then. This consequence seems troubling: we typically regard determinism as an open epistemic possibility, yet we do not tend to judge that had anything at all been different then the initial condition of the universe would have been different. The solution Lewis adopted was to permit small violations of law so as to maintain exact match over the vast majority of history. Necessarily indeterministic laws, as in quantum modal realism, offer a different solution: if the laws are necessarily indeterministic, then we simply do not need to worry about how counterfactual thought works under determinism. Since all Everett worlds are indeterministic, quantum modal realism faces no problem of counterfactuals in deterministic worlds.

Since he was committed to capturing intuitive counterfactual judgements even in the context of deterministic worlds, Lewis appealed to miracles: small, localized, violations of law. Quantum modal realism does not encompass any worlds containing violations of the fundamental laws; but (as we shall see in chapter 4) it does encompass plenty of worlds containing violations of non-fundamental laws. The Lewisian account of the closeness ordering can therefore be adapted to quantum modal realism by substituting suitable low-chance quantum events for genuine miracles. Lewis (1979/1986a) offers the following four criteria for settling a closeness ordering:

1. It is of the first importance to avoid big, widespread, diverse violations of law.
2. It is of the second importance to maximize the spatio-temporal region throughout which perfect match of particular fact prevails.
3. It is of the third importance to avoid even small, localized, simple violations of law.
4. It is of little or no importance to secure approximate similarity of particular fact, even in matters that concern us greatly.

Lewis (1979/1986a: 47–8)

These criteria do a reasonably good job of capturing many ordinary counterfactual judgements. They can be straightforwardly emulated—if desired—by quantum modal realists. The role of genuine violations of law is played, in a quantum modal realist implementation of the Lewisian closeness criteria, by what I will call large-scale *quantum miracles* and *thermodynamic miracles*. These events respect the fundamental laws, while being very low probability.[6] For the larger miracles that feature in clause 1, there will in general be large-scale quantum miracles that can be substituted for them.[7] For the small miracles that feature in clause 3 of the Lewisian criteria, there will in general be thermodynamic miracles that can be substituted for them.

In offering their substitutes for miracles, quantum modal realists take advantage of some distinctive features of the space of Everett worlds: the well-behaved objective chance measure covering the entire space, and the linearity of the Schrödinger equation. The combination of these two features ensures that there are Everett worlds involving phenomena of absolutely tiny objective probability: the sorts of 'quantum leaps' or (as I will call them) quantum miracles that might manifest in a teacup teleporting a foot to the left, in the spontaneous shattering of a diamond, or even in a miraculous reanimation of a corpse. Such miracles correspond to the 'tails' of the wavefunction as it spreads out throughout the space in which it lives. These tails are notorious problems for dynamical collapse approaches to quantum theory such as the GRW approach (Ghirardi et al. 1986), but Everettians may simply regard the wavefunction tails as an unproblematic consequence of quantum mechanics. For quantum modal realists, the tails also have significant potential metaphysical applications.

One important role of the wavefunction tails in quantum modal realism is to reduce the intuitive cost of accepting necessitarianism about laws of nature by underwriting an enormously expanded range of (albeit very low probability) deviant quantum possibilities. These scenarios are in general conceivable, and accordingly if quantum modal realism can accommodate them as possibilities it can recapture a broader class of instances of the familiar conceivability–possibility link. Our present concern, though, is the

[6] Quantum miracles have been discussed in the context of the Lewisian theory of counterfactuals by Hawthorne (2005) and Hájek (MS).

[7] At least, this is so if the miracle is metaphysically possible. The Lewisian framework anyway breaks down in the presence of counterpossibles, as discussed below.

role of the tails in accounting for counterfactual evaluation through featuring in a set of quantum modal realist closeness criteria.

Typically a counterfactual antecedent may be realized in a variety of ways, with a variety of corresponding objective chances. Consider the scenario in which all the molecules in the room end up located momentarily in the top-left corner. We can obtain that scenario by a brute-force quantum fluctuation that has the molecules spontaneously localize there. I call this sort of realization of the antecedent a large-scale quantum miracle. Alternatively, we can realize the same scenario more subtly by making small tweaks to the momentum of each molecule so that after a few more collisions they conspiratorially all end up travelling corner-wards, with the entropy of the situation tending to decrease. I call the latter sort of realization of the antecedent a thermodynamic miracle. Another nice example of a thermodynamic miracle, due to Hall (2011) in a review of Lange (2009), is a remarkable convergence of momenta of molecules in a body of water that causes jets of water to erupt up in precisely the right pattern to allow a person to walk on the water. Compared to the large-scale quantum miracle required to teleport directly all of the same water molecules repeatedly upwards out of the water, the chance of the process that gives the water molecules conspiratorial-looking trajectories will be relatively more probable. Thermodynamic miracles will typically be the highest-chance realizers of ordinary counterfactual antecedents.

Let us now apply these two kinds of miracles to adapt the Lewisian criteria to the quantum modal realist setting. The following criteria generate results broadly similar results to those generated by the Lewisian criteria:

1. It is of the first importance to avoid large-scale quantum miracles.
2. It is of the second importance to avoid thermodynamic miracles.
3. It is of the third importance to maximize perfect match with respect to the past light-cone of any events specified by the antecedent.
4. It is of little to no importance to secure approximate similarity of particular fact, even in matters that concern us greatly.

To assess ordinary counterfactuals, such as 'if I had not dropped the cup, it would not have shattered', we first check whether the antecedent can be implemented without any miracles at all. So we move to maximizing perfect match with respect to the past light cone of the event specified by the antecedent: how things were prior to my actual dropping of the cup. The difficulty now is that the histories with closest match in which I do not drop the cup involve some kind of quantum miracle or other: my body is in a state that makes dropping close to inevitable, but somewhere along the causal chain there is a quantum miracle and suddenly things are back on a non-dropping track. So we need what is referred to in the Lewisian semantics as the transition period: a 'smooth alteration' from the actual contents of the past light cone to one compatible with the antecedent.

Given the criteria proposed here, this transition period will be implemented in the closest worlds in the way which involves the least intrusive possible miracle: ideally with no miracle at all ('if the particle had been measured spin-up in the z direction...'), or (less ideally) with only a thermodynamic miracle ('if a gust of wind had spuriously triggered the detector', or (least ideally) with a large quantum miracle ('if I had teleported a foot to the left...'). What has been implemented through the modified Lewisian criteria is a procedure of balancing faithfulness to the actual contents of the past with the minimization of miraculousness in any transition from that past to the events specified in the antecedent.

Although the modified Lewisian closeness ordering, when plugged into a Stalnaker–Lewis semantics for counterfactuals, provides a plausible and relatively conservative account of counterfactuals, it is not obligatory for quantum modal realists to adopt it. Other theories of the semantics of counterfactuals may also be constructed using the flexible theoretical resources of the Everettian multiverse. Before moving on, I will outline a potential alternative approach to counterfactual semantics given directly in terms of the global quantum chance measure. On this alternative approach, counterfactual conditional assertions are assertions directly about the conditional chances of various possibilities. The guiding idea is that if A had been the case, B would have been the case if and only if the conditional chance of B given A is 1. If we had been on time, we would have caught the train if and only if (in context) the conditional chance of us catching the train given that we arrive on time is 1. This account would have to rely heavily on a contextualist theory of chance whereby in some ordinary contexts outcomes can often have chance 1; one such contextualist account is given by Handfield & Wilson (2014), and another by Hoefer (2007). Working out the details of this chance-based counterfactual semantics will have to await another time, but for now the point I wish to make is that quantum modal realism provides enough raw theoretical resources to build realist chance-based accounts of counterfactuals that eschew miracles of any kind.

The main message of this section so far has been that quantum modal realism provides rich theoretical resources that can be deployed to construct flexible semantic theories for counterfactuals. Crucially, the resulting theories promise to underwrite naturalistic explanations of the central role of counterfactuals in our cognitive lives. Accounts of the closeness ordering that draw on the global chance measure over the space of worlds are in a position to parlay the Everettian explanations of the knowability and practical relevance of chance that are described in chapter 3 into an account of the knowability and practical relevance of counterfactuals. Since Everettians are in a unique position to explain the knowability and practical relevance of chances, they are in a unique position to explain the usefulness of our employing a counterfactual conditional construction which conveys information about these chances. This promises a satisfying answer to the question concerning the aetiology of counterfactual discourse that has sometimes been developed in the form of an objection to the Stalnaker–Lewis semantics (see e.g. A. Wilson 2013b): why should creatures like us have adopted a counterfactual construction with a semantics like *that?* Quantum modal realists can respond; because it allows us to communicate facts about conditional chances, facts which may be directly relevant to our future prospects and hence can guide rational action. Lewisian modal realists cannot sustain the same response, at least not in full generality.

Counterpossible counterfactuals are the elephants in the room which now need addressing. Were I to encounter a round square, I should be very surprised. If there were no numbers, there would be no primes. If wishes were horses, beggars would ride. These apparently robust counterpossible judgements form the basis of one of the most potent objections to the Stalnaker–Lewis semantics: that semantic theory makes no non-trivial distinctions in truth-value amongst counterfactuals with metaphysically impossible antecedents. Since all possibilities are such that no impossibility obtains, there are no distinctions amongst possibilities available to be exploited by the semantics. There are never any closest antecedent-worlds because there are never any antecedent-worlds at all. This problem has long been recognized (e.g. by Stalnaker 1968, Lewis 1973: 24) but has never, I think, been adequately defused. Field (1989) gives some compelling cases of non-trivial counterpossibles in mathematics, which have not been convincingly rebutted in the literature.

Friends of the Stalnaker–Lewis approach to counterfactuals have typically tried to explain away our different reactions to different counterpossibles in pragmatic terms. For example, Lewis (1973) suggests that '[counterpossibles] do not have to be made false by a correct account of truth-conditions; they can be truths which (for good conversational reasons) it would always be pointless to assert' (Lewis 1973: 25). Employing a related strategy, Stalnaker (1984) aims to model our attitudes to counterpossibles using metalinguistic resources: we adopt non-trivial attitudes to the proposition that the counterpossible in question is in fact a counterpossible, and these non-trivial attitudes then go proxy for our attitude to the counterpossible itself. It is safe to say that neither of these defensive strategies has proven particularly popular; however, in recent years Timothy Williamson has given counterpossible triviality a renewed and more persuasive defence. Williamson (2007) defends counterpossible triviality instead by going on the offensive, arguing that 'the logic of quantifiers was confused and retarded for centuries by unwillingness to recognize vacuously true existential generalizations; we should not allow the logic of counterfactuals to be similarly confused by unwillingness to recognize vacuously true counterpossibles' (Williamson 2007: 175). A more elaborated response is in Williamson (forthcoming), which explains our inclination to regard certain counterpossibles as false in terms of the operation of some convenient heuristics, which are normally helpful in counterfactual reasoning but which lead us astray in the abstruse contexts of counterpossible reasoning.

Philosophers more impressed by the apparent cogency of (at least some) counterpossible reasoning have developed strategies of two rather different kinds for dealing with them. The first strategy, associated in particular with Nolan (1997), Goodman (2004), Priest (2005), and Jago (2014), extends the Stalnaker–Lewis semantics by quantifying not over possible worlds but over worlds in a more general sense. According to these authors, some worlds are possible and some worlds are impossible. They typically reject what Nolan has called the *strangeness of impossibility* condition (Nolan 1997), which specifies that all possible worlds are closer to the actual world than is any impossible world. The second strategy, associated with authors such as Martin (2008), Borghini and Williams (2008), Jacobs (2010), and Vetter (2015), aims to do away with possible worlds altogether and to replace the Stalnaker–Lewis approach with an altogether different semantics based on modally rich properties of individuals, such as essences or dispositions.

The first strategy, which appeals to impossible worlds, can be adapted straightforwardly by the quantum modal realist if they wish. This is because (non-concrete) impossible worlds can be constructed in a straightforward way using the resources of quantum modal realism: this ersatz construction procedure will be described in more detail in section 1.4. The second strategy, which abjures worlds altogether, is part of a metaphysical vision very different from quantum modal realism. Even so, the second strategy is potentially available to Everettians in virtue of the rich pattern of conditional chances that is a part of quantum modal realism: if we could give an account of dispositionality directly in terms of chances, then perhaps we could adapt the second strategy to the Everettian context.

My own inclination is to reject both strategies, and simply to embrace counterpossible triviality. Here I am motivated by a desire to maintain the guiding Stalnaker–Lewis thought with which we began this section: the thought that counterfactual conditional assertions are, in the final analysis, *about* how things stand with respect to genuine alternative possibilities. By accepting non-trivial counterfactuals, one gives up on this thought. Something remains—the thought that counterfactuals are about how things stand with respect to describable scenarios, perhaps—but something is also lost. We retain a simpler and more compelling connection between modality and counterfactuals if we follow Stalnaker, Lewis, and Williamson in endorsing counterpossible triviality. However, it is important to see that counterpossible triviality is not forced on us by quantum modal realism; in the end, quantum modal realism is a metaphysical framework and the semantics of counterfactuals is a project in natural language semantics.

Even though limitations with respect to counterpossibles are a general characteristic of Stalnaker–Lewis approaches to counterfactuals, it might be thought that the problem is especially acute for quantum modal realists. The two quantum modal realist approaches to counterfactuals that I have sketched in this chapter both count counterfactuals with physically impossible antecedents as trivially true.[8] If there are no worlds where the antecedent holds, then the consequent is true at all antecedent-worlds. And if an event is impossible, it has chance zero conditional on any possible event. In combination with the modal necessitarianism incorporated into quantum modal realism, this prevents either of my proposed accounts from vindicating the widespread intuition that there are some false counterlegals (counterfactuals with antecedents which are logically consistent but which violate the laws of nature). This may seem like a difficulty for quantum modal realism, since such counterfactuals seem to be assessed frequently in scientific reasoning.[9]

[8] At the cost of slightly complicating the formulation of the account, it would be possible to take all counterpossibles as trivially false rather than trivially true. Not much hangs on this; the crucial point is that the account does not permit variation in truth-values of different counterpossibles.

[9] Tan (forthcoming) provides a comprehensive survey of counterlegal reasoning in scientific practice.

My response to the challenge of counterlegals is twofold. Firstly: a large proportion of apparent counterlegals—those with antecedents that violate only non-fundamental laws of nature—can indeed be made non-trivial sense of in quantum modal realism. (See section 4.6 for more details.) So the difficulty affects antecedents that violate fundamental laws. Secondly, and more generally: the triviality of counterlegals is a feature, not a bug, of quantum modal realism. Since the fundamental laws of nature are necessary according to quantum modal realism, there is nothing non-trivial to be said about genuine possibilities in which these laws fail to hold.

Counterlegals do feature in our epistemic practices, but so do other counterpossible counterfactuals that cannot be handled by any mainstream realist theory of counterfactuals. Orthodox Stalnaker–Lewis approaches already fail to deliver intuitively correct verdicts for mathematical counterpossibles, which (as Field emphasizes) seem to play a role in mathematical discovery. Likewise, they cannot allow for non-trivial counterfactuals with antecedents specifying that non-classical logics hold, of the sort we might use in debates over which logic is correct. Most tellingly in the present context, they cannot allow for non-trivial counterfactual reasoning about which theory of modality is correct. According to Lewis, for example, any claims about what would be the case if modal realism were false are 'nonsense, intelligible only if taken as vacuous' (Lewis 1986b). The failure of quantum modal realism to make sense of counterfactual reasoning about scenarios it counts as genuinely impossible is of a piece with these well-known features of the possible-worlds semantics. While quantum modal realism counts some scenarios as impossible that are possible according to Lewisian modal realism, both theories require counterpossible reasoning—including reasoning that may play important epistemic roles—to be handled differently from non-counterpossible counterfactual reasoning. It is no deep cost of quantum modal realism that it moves counterlegal reasoning into the former category. This theme will be elaborated in section 1.4; meanwhile, it is time to sum up the argument of this section.

Quantum modal realism can support a variety of approaches to the semantics of counterpossible conditionals, including versions of the justly popular Stalnaker–Lewis possible-worlds semantics that replace violations of laws of nature with low probability but physically possible quantum events. For the purposes of my overall project in this book, we need not settle on a specific approach to counterfactuals; we need only observe that the global chance measure defined over all Everett worlds provides quantum modal realists with strictly richer theoretical resources to work with than we have on the Lewisian picture. Most importantly, semantic theories of counterfactuals that give a significant role to Everettian chances will allow for a unique explanation of the knowability and practical relevance of counterfactuals. All of this adds up to a significant line of argument in favour of quantum modal realism.

1.4 Quantum Modal Realism at Work: Content

One of the central applications of Lewisian possible worlds is to characterize the content of our mental and linguistic representations. In this section I will show how the Lewisian modal approach to representational content can be carried over in full generality to quantum modal realism. While the space of Everett worlds is more restricted than the space of Lewis worlds—there are no Everett worlds in which the fundamental laws are non-quantum, for example—this is no deep threat to the adequacy of an Everettian analysis of content. This is because the Lewisian account already can—and already does—routinely handle contents corresponding to genuinely impossible scenarios in an indirect pragmatic way, alongside its direct semantic treatment of contents corresponding to genuinely possible scenarios. Everettians can likewise draw a sharp distinction between semantically and pragmatically communicated representational content, though they will draw this distinction in a slightly different place.

The modality that is at the centre of quantum modal realism is an objective modality: what is and is not possible is independent of thought and language. But this does not mean that Everettians cannot also make clear sense of modalities that do depend on thought and language, including epistemic and conceptual modalities. Indeed, I will argue that quantum modal realism gives us rich resources with which to do so, resources that closely mirror those of Lewisian modal realism. Later in this section I will characterize the Everettian modality as an objective modality that is the *norm* of subjective modality: subjective modal thought aims at the objective modal facts, and its rational role is to be explained in terms of the objective modal facts. Thus metaphysical possibility stands to epistemic possibility as chance stands to credence. But before introducing this account in detail, we must get clear on what the various non-objective modalities are like, and on how they derive from the space of metaphysical possibilities constituted by the Everettian multiverse.

To make room for non-objective modalities, we need to move beyond the coarse-grained conception of representational content that we have dealt with thus far. The account of propositions as sets of Everett worlds is adequate for contingent propositions that can serve as the bearers of objective chance. It is thereby adequate to draw all the real distinctions between objective possibilities that are out there in the world. However, it is inadequate to capture all of the distinctions drawn by our imperfect forms of communication. Limited creatures that we are, we draw some distinctions that do not correspond to any differences in reality, and we fail to draw some distinctions that do correspond to differences in reality. To model what we mean when we express distinctions that do not correspond to any real difference, we need more fine-grained contents. We need what I call *scenarios* in addition to possible worlds.

We will begin with the hardest case: mathematics, and other domains of necessary truth. The account of propositions as sets of worlds, incorporated into quantum modal realism, is only adequate for contingent propositions. Any necessary proposition corresponds to the same set of worlds—all of them—and so the coarse-grained account of propositions is unable to distinguish them. Accordingly it cannot account for the difference in content of different logical or mathematical truths, and it cannot distinguish between the content of any logical or mathematical truth and the content of fundamental laws of nature (see section 4.8 for additional discussion of this point). We need more finely-grained contents. Lewis (1986b) recognized the need for fine-grained representational contents, and offered a construction strategy for such contents founded on the rich resources of his plenitude of possibilia. Chalmers and others have since developed more sophisticated versions of the same general strategy, though since most such authors abjure modal realism these versions tend to lack the solid foundations of the Lewisian account.

In the spirit of the modal realist's constructive metaphysical stance, I will describe a general procedure for constructing fine-grained contents out of the coarse-grained materials provided by quantum modal realism. One desirable feature of a convincing hyperintensional notion of content, that goes beyond the real physical differences in the world, is that it be modelled on the intensional notion of content that corresponds to those real physical differences. That feature is desirable because it permits a plausible account of how hyperintensional thought arose, as intertwined with intensional thought. Since on the quantum modal realist account intensional thought is fundamentally thought about alternative possible worlds and the patterns therein, it is natural to model hyperintensional thought as fundamentally thought about alternative scenarios and the patterns therein, whether or not these scenarios correspond to genuine possibilities. Thus in my proposed account of hyperintensionality we have the same basic operations of thought—quantification, cross-world correlation via counterpart relations, closest-worlds conditional semantics—but operating on different types of entities: scenarios as opposed to Everett worlds.

Whether for the reason just sketched or not, a scenario-based approach to hyperintensional content currently dominates the literature. Different authors tend to disagree on the nature of scenarios and of the logics appropriate to them more than on the core idea of quantifying over something that is scenario-like. For example, we have Priest's *open worlds* (Priest 2005), Chalmers's *epistemic scenarios* (Chalmers 2011), and the *impossible worlds* of Nolan, Goodman, Jago, and Berto (Nolan 1997, Goodman 2004, Jago 2014, Berto and Jago 2019). All of these authors give constructions of a space of scenarios that permits distinctions finer than the distinctions that can be made with respect to objective modal space.

There has been resistance in some quarters to impossible worlds, related to the scepticism about non-trivial semantics for counter-possible counterfactuals that is expressed by Stalnaker (1996) and Williamson (2007). Still, these authors do not deny the need for a non-trivial epistemology of necessary truths such as those of logic and mathematics. Stalnaker, for example, endorses a meta-linguistic story about the epistemology of mathematics (Stalnaker 1984) and Williamson's epistemology of the non-contingent is bound up with his distinctive knowledge-first externalism (Williamson 2007). What I will now outline is one, relatively neat, specific way of constructing the needed contents; it is certainly not the only option available to quantum modal realists, and it may not be the best, but it will serve as proof of concept.

Lewis (1986b: 30–2) offered a sketch of a treatment of disagreement over necessary truths, in particular over the laws of logic. It involves attributing multiple inconsistent contents to an agent's belief state, the extent of the multiplicity representing the extent of inconsistency (or, as Lewis, puts it, of 'equivocation') in the corresponding belief state. Greg Restall has since shown how to generalize this strategy in a very elegant way to a full notion of hyperintensional content (Restall 1996, 1997). Restall constructs generalized *worlds* as nonempty sets of possible worlds; whereas we can think of a possible world as a way things could be, worlds in the generalized sense are ways things either could be or could not be. The sets of possible worlds that correspond to ways that things could not be are inconsistent, in the following sense: they represent things inconsistently, as being the way their first possible world member represents things *and* as being the way their second possible world member represents things *and*... and so on. Our purposes here will not require us to delve into the details. Armed with some construction or other of generalized worlds, we are in a position to characterize the full space of hyperintensional content in terms of sets of generalized worlds.

So what *are* propositions in the quantum modal realist picture? Are they sets of epistemic scenarios or sets of Everett worlds? This dispute is largely verbal: quantum modal realists can characterize both sorts of proposition and use them both in their appropriate roles. As Lewis put it in the context of his own modal realism: 'there is no contest between structured and unstructured versions of the properties, relations, and propositions. Given the combined resources of modal realism and set theory, we have both versions' (Lewis 1986b: 59). Quantum modal realism gives us, in its sets of Everett worlds, coarse-grained propositions with a distinctively privileged metaphysical status. These propositions are the primary bearers of objective chances and the loci of emergent contingency. More complex constructions out of Everett worlds and individuals within them do not supplant these basic contingent propositions, and they cannot play the same core theoretical roles with respect to chance and contingency. Still, the more complex constructions have a considerable role to play in accounting for our broader modal thought. Such a two-tier conception of modality has already been defended in the non-Everettian context by Berto (2010).

Once we have scenarios in our quantum modal realist framework, we can apply them beyond simple attributions of epistemic or conceptual possibility and necessity. Ordinary modal thought and talk contains a number of modal verbs and adjectives; the operators 'possibly' and 'necessarily' that featured heavily in section 1.2 are in fact rather unusual pieces of vocabulary in ordinary English. Much more common are instances of the modal verbs 'might', 'may', 'could', and 'must'. Even the briefest attention to the way in which we use these modal verbs reveals that frequently—if not always, as many have suspected of 'might' and 'may'—these modal verbs have 'epistemic' uses characterized, roughly, by the relation of the modified clause to some aspect of the speaker's state of mind.[10] For example, the phrase 'it is

[10] Relativists about epistemic modal verbs generally allege also that these epistemic uses are characterized by the relation of the modified clause to some aspect of the assessor, or of the

possible that' is nearly always used in an epistemic sense outside philosophy departments. There is a substantial and growing literature surrounding these 'epistemic modals'. (See Egan & Weatherson 2011 for a flavour of this literature.) For present purposes, we do not need to get involved with the details of the various proposals. Whether epistemic modals express credences or state facts about bodies of information or operate on the conversational context need not detain us; but we can note that none of them take the business of epistemic modals to be to state facts about the objective possibilities. This different communicative function sets epistemic modals as a category apart from what I will call 'objective modals'; uses of modal verbs like 'can', and 'must' that really are in the business of stating facts about objective possibilities and necessities.

We can draw the threads of the above discussion together into a substantive distinction between *objective modality* and *subjective modality.* I propose giving the former, but not the latter, a direct analysis in terms of Everett worlds and the chance distribution over them. To refine the objective/subjective distinction a little further, we should distinguish *alethic* from *non-alethic* applications of modal thought. Such applications include, but are not limited to, deontic modality (what is morally permissible), social or legal modality (what is compatible with prevailing social or legal norms), and bouletic modality (what is compatible with some agent's desires). All of these non-alethic modalities I propose to situate on the subjective side of the objective/subjective distinction. Subjective modality takes many forms, corresponding to all the different mental attitudes we could have towards some content; but in every case the content in question will be given in terms of scenarios, not Everett worlds.

Focus now on the distinction between objective alethic modality and subjective alethic modality, setting aside subjective non-alethic modalities. On the one side, there is objective alethic modality: a set of facts expressible via the 'possibility' and 'necessity' operators as well as with modal verbs like 'can' and 'must'. These objective modal facts depend on features of the Everettian multiverse, including the relative weights of Everett worlds; while they may have an indexical component, they do not constitutively depend on any aspects of agents. On the other side, there is subjective alethic modality, which does depend constitutively on the various cognitive relations we might bear towards scenarios. A central distinction within subjective alethic modality is

context of assessment. See MacFarlane (2014) for more discussion. Relatedly, Stalnaker (2014) argues that the content of epistemic modals depends on the conversational common ground.

between epistemic possibility (compatibility with an agent's knowledge) and doxastic possibility (compatibility with an agent's beliefs). Although questions about the relation between these are crucial for epistemology, for present purposes I will set these questions aside.

My proposal is that subjective alethic modality *aims at* objective alethic modality. This way of conceiving of the relation between objective and subjective modality is a close relative to the familiar thought that subjective credence *aims at* match with objective chance, as captured by the Principal Principle of Lewis (1980/1983). Objective alethic modality thus stands to subjective alethic modality as chance stands to credence:

Modal Principal Principle: Match with the objective alethic modal facts is the norm of subjective alethic modal thinking.

This formulation is intentionally open-ended, so as to be compatible with a variety of approaches to subjective alethic modality. But I hope the main idea is clear; and it is easy enough to see how the Modal Principal Principle plays out in simple examples. Consider a rational agent A of the kind Lewis held to be bound by the Principal Principle. If A has credence 1 that p is objectively necessary, then A should regard p as epistemically necessary: A should assent to 'it must be that A' and deny 'it might not be that A'. If A has credence 1 that p is objectively possible, then A should regard p as epistemically possible: A should assent to 'it might be that A' and deny 'it must not be that A'.

There are familiar limits of the formulation of the chance-credence connection in terms of a constraint on rational initial credence functions, as in the Principal Principle. Such formulations fail to entail anything directly about the less-than-perfectly-rational agents in which we are typically interested; at most, they tell us about how we should behave or about how we would behave if we were better in certain respects than we actually are. These limits are especially troublesome when it comes to the epistemology of a priori disciplines such as logic and mathematics. Since logical truths have chance 1 and logical falsehoods have chance 0, the Principal Principle tells us that we should be certain in all the logical truths and that we should assign no credence to all the logical falsehoods. To obtain an adequate epistemology of a priori subject matters, we need to go beyond the space of metaphysically possible worlds that quantum modal realism offers. I will not explore this difficult terrain any further, although I suspect that the construction of scenarios described above may have a useful role to play.

This section has aimed to convince the reader that quantum modal realism provides the sorts of rich theoretical resources for constructive theorizing about content that Lewis promised as a benefit of his modal realism. In section 1.5, I turn to the work that quantum modal realism can do in philosophical theorizing about properties.

1.5 Quantum Modal Realism at Work: Properties

What are properties? To a first approximation, the theoretical role of properties is to be whatever it is that different individuals have in common when they resemble one another. Any entity that can play the property role thus characterized must allow us to pick out a set of individuals at a given world: those individuals which have that property. Each property had by anything actual corresponds to a set of individuals at the actual world: these actual individuals resemble one another in virtue of having the property in question.

It would be nice to be able to stop here, and simply identify properties with the set of their actual instances (or some variation thereof). For example, we could identify charge with the set of charged things and mass with the set of massy things. The flaw in this plan is familiar and fatal: there are actual cases of accidentally coextensive but distinct properties. The classic example concerns hearts and kidneys; it is supposed that as a matter of fact every animal with a heart also has a kidney, but that this need not have been so.[11] In order that being hearted and being kidneyed should not come out as the same property, we need to move beyond the simple extensional analysis of properties as sets of instances.

Intuitively, what is wrong with the simple extensional analysis is that while all actual hearted creatures are (let us suppose) kidneyed, it is possible for something to have a heart but lack a kidney or vice versa. For modal realists, Everettian and Lewisian alike, possibility claims involve quantification over the individuals throughout some region of modal space. So a natural extension to the extensional programme accordingly quantifies not over actual individuals but over actual and possible individuals. The property of having a heart is the set of all actual and possible hearted creatures; since the set of possible hearted creatures is distinct from the set of possible kidneyed creatures, the properties come out distinct as desired.

[11] As it happens, this is wholly false: numerous invertebrates have hearts but no kidneys.

The motivation for the modal realist theory of properties is not its intuitive plausibility; an analysis of properties as sets of any kind is not obvious, to say the least. Rather, the plausibility of the modal realist analysis derives (as in the case of the modal realist approach to counterfactuals) from its simplicity and explanatory power. In the same way that modal realists can treat counterfactual conditional assertions as directly *about* what goes on at other possible worlds, modal realists can treat property-ascriptions as directly about what unifies individuals with those other (actual and possible) individuals that they objectively physically resemble.

The intensional theory of properties is the core of quantum modal realism's approach to properties. But as Everettians need to generalize their basic intensional theory of content to allow for analysis of the mental and linguistic representations of imperfect agents, so they need to generalize their basic intensional theory of properties to allow for analysis of the predicative portion of those representations. When Johann Becher hypothesized that burning objects released their phlogiston, he posited properties—*phlogisticated* and *dephlogisticated*—that do not in fact contribute towards any measures of objective resemblance between objects. Since there is no phlogiston, there is neither any phlogisticated air nor any dephlogisticated air. Nonetheless, we need to characterize what was going on when Becher formulated his theory. To do this without giving up on the prospect of a straightforward referential semantics for Becher's language, we need to come up with a semantic value for the term 'phlogisticated'.

As will be becoming a familiar refrain, quantum modal realism provides rich theoretical resources for coming up with hyperintensional theories of content. Applying the general approaches to content of section 1.4 to the semantic content of predicates in particular, we can characterize hyperintensional properties via a dog-leg through scenarios. Scenarios contain all logically coherent (intended as a very weak constraint indeed) combinations of terms in a language. Accordingly, they involve the attributions of predicates built up through arbitrary—though logically coherent—combinations of intensional predicates, including several combinations which are impossible. Since the impossible scenarios do not collapse into one and the same scenario in the same way in which all impossible coarse-grained propositions collapse into one and the same coarse-grained proposition (the empty set of Everett worlds), the differences between impossible scenarios allow for characterization of hyperintensional properties. Such constructions characterize ways that a thing might be thought to be, as opposed to ways that a thing could in fact be. The properties that are out there in the world are still characterized directly in terms of sets of quantum possibilia.

Are intensional properties or hyperintensional properties *really* properties? As with the nature of propositions, modal realists need not regard this as a significant question. Given the basic ontological resources of quantum modal realism, there are intensional properties (sets of individuals in Everett worlds) and there are hyperintensional properties (semantic values of the predicates that form part of epistemic scenarios, which themselves are constructed according to something like Restall's procedure). We can theorize with both kinds of property, another benefit of the Everettian paradise of possibilia.

1.6 Isolation

If quantum modal realists are to countenance an enormous plurality of Everett worlds, and to ascribe to them the metaphysical role of genuine alternative possibilities, we owe a story about the individuation of Everett worlds. Under what conditions are two entities part of the same Everett world, and under what conditions are they parts of different worlds? My full answer to these questions draws on the physical process of decoherence. In section 2.3 I shall sketch the physics of decoherence, and show how it is used by Everettians to identify a robust structure of emergent worlds. In this section, I consider two more familiar individuation criteria for worlds—a spatiotemporal criterion and a causal criterion—and conclude that neither of them need be taken as axiomatic by quantum modal realists. Lewis took the spatiotemporal criterion for demarcation of worlds as axiomatic, and derived causal isolation from it via his counterfactual theory of causation. I shall suggest that quantum modal realists should reverse this approach, by deriving spatiotemporal isolation from causal isolation (at least in domains where spacetime is well defined). Causal isolation, in turn, can be naturalized through being underwritten by the decoherence process.

Consider first the spatiotemporal world demarcation criterion incorporated into Lewisian modal realism, according to which two events are worldmates if and only if they bear some spatiotemporal relation (or some analogous relation) to one another. This criterion makes worlds into maximal webs of spatiotemporally related events. A world is an island universe.[12]

[12] This feature gives rise to some familiar objections to Lewisian modal realism based around the possibility of distinct island universes within a single world (Bigelow & Pargetter 1987; Bricker 2001). The treatment of advanced modalizing discussed in section 1.2 above undercuts these objections.

We might wonder, especially in the light of ongoing exploration of approaches to quantum gravity in which our ordinary space and time are non-fundamental, whether it is appropriate to build spatiotemporal relations into the notion of a world from the outset. Lewis was alive to this concern, speculating there could be worlds with alien perfectly natural relations playing analogous roles to the roles that spatiotemporal relations play in the actual world. He therefore generalized the spatiotemporal criterion into a more general criterion that makes worlds into maximal webs of events related by *analogously spatiotemporal* perfectly natural relations. Through this move Lewis intended to exclude relations of the kind that could hold between worlds. Even if, for example, relations of same-chargedness or same-numerosity might hold between entities in different worlds, these relations are not analogously spatiotemporal and hence not suited to demarcating one world from another. However, until we are told more about what makes for being analogously spatiotemporal, the Lewisian spatiotemporal demarcation criterion remains underspecified.

Whereas Lewis seems to have been motivated by the apparent conceivability of worlds without space or time, quantum modal realists who reject the conceivability–possibility link will be unmoved by this sort of reason to endorse analogously spatiotemporal relations. Quantum modal realists do however have to contend with the prospect of emergent space and time in quantum gravity and hence with the possibility that spatiotemporal structure varies from world to world; some such possibilities will be discussed in more detail in chapter 4. There is no guarantee from current physics that all Everett worlds will be spatiotemporal, or even analogously spatiotemporal—whatever precisely that might amount to. It is therefore worth looking elsewhere for a world demarcation criterion better suited to quantum modal realism.

Consider next the causal world demarcation criterion. According to the causal criterion, two events are worldmates if and only if they are connected through a chain of causal relations—if one of the events directly or indirectly causes the other, if both events are causal descendants of some third event, or if there is some third event of which both are causes. This criterion makes worlds into maximal networks of causal relations. A world is a complete causal web.

Lewis accepted that worlds are causally isolated, but this was not a basic assumption of his modal realism; instead he derived causal isolation from the analogously spatiotemporal criterion via his counterfactual theory of causation. No event in one world counterfactually depends on any event in any other world, so if causation is a matter of patterns of counterfactual dependence then no event in any world causally depends on any event in any other world. Likewise, a counterfactual account of causation, in the presence of an account of counterfactuals as statements about how things stand at other Everett worlds, guarantees that Everett worlds are causally isolated. What vindicates our treating Everett worlds as alternative possibilities is decoherence, and hence it is decoherence which is ultimately responsible for the causal isolation of the Everett worlds. Causal isolation is derived rather than axiomatic within quantum modal realism, but it is derived in a different way from that of Lewisian modal realism.

In chapter 2, I shall give more details of the process of decoherence, and how it accounts for the causal isolation of Everett worlds. For now, I shall just describe the overall picture I have in mind. Everett worlds do not causally interact with one another. Still, we can explain the outcome of quantum experiments in terms of the existence of nearby worlds. It is because of the presence of other Everett worlds that things are as they are in our own Everett world. Individual histories are components of a larger emergent structure, and so they can be explained non-causally by appeal to global features of that structure. By providing information about nearby Everett worlds, we provide an account of the larger structure of which our own world is a component. This of course entails that not all our scientific explanations are causal explanations: there are some genuine non-causal explanations of occurrent facts.

Consider again the metaphor of a jigsaw puzzle from section 1.1. Why is some particular piece the shape that it is? Because it fits into the space left by the pieces around it. The shapes of the surrounding pieces do not cause our piece to be the shape that it is (that is the role of the machine that cut the jigsaw), but the shapes of the surrounding pieces nonetheless provide a non-causal explanation of our piece's shape. So it is, I suggest, with Everett worlds: there is instead a single common source of the features of the various Everett worlds, but the features of nearby worlds nonetheless provide a non-causal explanation of the features of our world. The analogy, of course, breaks down in that the jigsaw pieces have their shapes as the result of a single common cause (the cutting machine) while the common explanatory factor in the quantum-mechanical case (the fundamental quantum state) is not a common cause but instead is something like a *common ground*.[13]

[13] See Schaffer and Ismael (forthcoming) for an illuminating treatment of common ground explanations and of their application to quantum entanglement.

The aims of this section have been modest: to distinguish spatiotemporal from causal isolation of worlds, to argue that both of these are derivative within the context of quantum modal realism, and to defer further discussion of how decoherence individuates Everett worlds to section 2.2. Like Lewis worlds, Everett worlds are spatiotemporally isolated and causally isolated, but they are individuated by physics rather than by any metaphysical principle.

1.7 Concreteness

Lewis (1986b) makes much of the difficulty of understanding what is meant by 'concrete' and 'abstract' in philosophical discourse. This is primarily because he is concerned to resist objections to his modal realism on the basis that it makes possible worlds concrete when they ought to be abstract. The same objections apply to quantum modal realism, but I regard them as completely unpersuasive; I suspect the influence of Kripke made them seem stronger than they were at the time that Lewis was writing.

There is no pre-theoretic datum that possibilities are concrete or that they are abstract—indeed, 'concrete' and 'abstract' are technical terms of metaphysics, so there are no pre-theoretic data about anything at all being concrete or abstract. I accordingly share many of Lewis's misgivings about the abstract–concrete distinction. Still, like Lewis, I would like to provide as much insight as possible into what the worlds of quantum modal realism are like. My preferred way to do this is to say that Everett worlds are worlds like this one, just more things of the same general sort. If the actual world is concrete—as I think it is, under any reasonable precisification of that technical term—then all worlds are concrete. Quantum modal realists may reasonably leave it at that and move on to more pressing matters, such as mapping the contours and limits of modal reality. That is the task of section 1.8.

1.8 Plenitude

Quantum modal realism tells us what possibility is. Unsupplemented, it does not tell us what things are possible. For modal realists of any kind, modality is fully objective and mind-independent: possibility and necessity are discovered, not invented. There is therefore no reason for modal realists to presume that we have perfectly accurate or complete epistemic access to the facts of modality. As Lewis puts it, on some questions:

> there seems to be no way at all of fixing our modal opinions, and we just have to confess our irremediable ignorance... certainly we are not entitled just to make the truth be one way or the other by declaration. Whatever the truth may be, it isn't up to us. Lewis (1986b: 114)

Modal realists hold that possible worlds exist prior to and independently of any of our representational practices. Partial agnosticism about the precise extent of genuine possibility is the only proper epistemic attitude to adopt once we have fully internalized a modal realist perspective. While realism about modality allows us to explain its relevance and its partial knowability, realism also puts some aspects of modality potentially beyond our ken. But this is exactly what we should expect, given the subsumption of modal knowledge under our (necessarily limited) scientific knowledge. Quantum modal realists should therefore embrace humility with respect to the exact contents of modal space.

It is one thing to decline to pass judgement on certain non-obvious claims concerning possibility; it is quite another to pass judgements that are incorrect. The most pressing concern about quantum modal realism for many readers will be the concern that it might be extensionally incorrect—it might simply get the modal facts wrong. For example, quantum modal realism entails that it is absolutely impossible for a massive object to accelerate past the speed of light, not just impossible with respect to some restricted space of possibility. And it is widely assumed, at least within the community of analytic metaphysicians, that breaking the light barrier in this way is genuinely possible; hence quantum modal realism is revisionary of ordinary opinion on this point. To assess how seriously we need to take this general concern, we shall need to delve deeper into modal epistemology.

The epistemology of modal truths has historically been found deeply mysterious. Whether we take knowledge of counterfactuals or knowledge of possibility and necessity to be primary, the process by which we acquire it seems opaque. The most prominent suggestions involve conceivability (e.g. Chalmers 2002), knowledge of our own concepts (e.g. Bealer 1996), or knowledge of essences (e.g. Fine 1994). However, these suggestions face considerable difficulties and—in my view—they only retain what popularity they have because of the perceived lack of any credible alternative.

In quantum modal realism, modal epistemology is entirely subsumed into general scientific epistemology. When we discover—experimentally or theoretically—that some outcome of some process has a non-zero objective chance, then we can immediately infer that there is a genuine possibility corresponding to it. For example, when we discover that uranium-235 is fissile, we *ipso facto* discover that there is an Everett world in which any given molecule of uranium-235 spontaneously decays within the next second of its existence. There is no further role for conceiving of things, or for interrogating our own concepts, or for a priori insight into essences. The move from chance to possibility does rely essentially on the core components of quantum theory, in particular on the Schrödinger equation. Once we identify the relevant features of the wavefunction of a uranium-235 atom which allow it to decay (i.e. the existence of a stable lower-energy atomic state) the Schrödinger equation provides assurance that there is some Everett world according to which an atom of that kind decays. It ensures that there are no gaps in the space of quantum possibilities.

The epistemological procedure described in the previous paragraph will work in all cases where the systems of interest are sufficiently decohered. Before we can even attribute a non-trivial chance to an outcome in EQM, however, we need to be assured that we have a suitably well-defined macroscopic outcome to which the chance can be assigned. Modelling an electron in an otherwise empty classical spacetime, for example, gives a wavepacket which spreads out steadily throughout space and which does not correspond to any determinate macroscopic outcomes. Without determinate outcomes that can be the bearers of objective chance, there are no decoherent histories describing the system and hence there are no things that quantum modal realists count as metaphysically possible worlds. Hence, to establish that a configuration of particles is genuinely possible, it is not enough to establish that the particles' system's wavefunction spreads out into a region of configuration space corresponding to that particular configuration. It is also necessary to establish that the wavefunction can be appropriately decomposed into a decoherent history space, as characterized in section 2.3. Still, this can in general be done for systems of macroscopic size, and hence quantum modal realists are able to recapture all the possibilities for macroscopic systems to which we have scientific reason to assign non-zero chance.

Is there any role left for conceivability in an Everettian modal epistemology? Conceivability looks like an objective alethic modal property; the property of possibly being conceived. Given the Everettian analysis of objective alethic modality, this property is analysed as the (extrinsic) property of being conceived in some Everett world. Thus conceivability is a perfectly respectable and naturalistic property; we are not left in the peculiar situation where a modal property (being possibly conceived) was supposed to hold the key to the fully general epistemology of modal properties. The potential circularity is broken in quantum modal realism by the openness of questions about conceivability to empirical test; there are ways of finding out whether something is in fact genuinely possibly conceived that do not amount to judging whether it is possible to conceive of its being conceived. Once we have this sort of naturalistic analysis of conceivability on the table, a further interesting option is analysing epistemic possibility in terms of conceivability, so that epistemic possibility is co-conceivability with what one knows. I will leave further exploration of these matters to another occasion.

One important consequence of quantum modal realism's reconceiving of conceivability is that it enables a powerful response to the formidable *modal arguments* for mind/matter dualism. These arguments, developed inter alia by Descartes (1641), Kripke (1980), and Chalmers (1996), derive anti-physicalist conclusions from the premise that mind-matter connections can be conceived not to hold. By undermining the inference from conceivability to possibility, and replacing it with an alternative naturalistic modal epistemology, quantum modal realism provides a one-size-fits-all response to modal arguments against physicalism (although, of course, this still leaves open which theory of consciousness is correct). The most prominent anti-physicalist arguments in the recent literature are therefore undermined at a stroke by quantum modal realism.

In the account of modal epistemology I have offered on behalf of quantum modal realists, the Schrödinger equation plays the role of what Lewis called a *principle of plenitude*: it ensures that there are no arbitrary 'missing possibilities'. A number of realist theories of the nature of modality, including Lewisian modal realism, appeal to some version of a principle of recombination to play the role of a principle of plenitude. We can think of recombination principles as material conditionals of the following form: for some special facts f: if it is possible that f_m and it is possible that f_n then it is possible that f_m *&* f_n. Lewis's informal 'patchwork principle' (Lewis 1986b) and the more rigorous treatment in Armstrong (1989) are familiar examples of recombination principles, but some such principle features in most views of possible worlds as real structured entities.

Surprisingly little attention is usually paid to the epistemic status of recombination principles. They are paradigm mysterious examples of the putative synthetic a priori—highly substantive truths about the nature of modal reality, our way to knowledge of which we are somehow supposed to be able to reason. Now that transcendental idealism is out of fashion, and conventionalism has had its day, the most plausible treatment of the epistemic status of putative synthetic a priori truths is the Quinean holistic epistemology outlined in section 0.4 of the introduction: such putative truths are justified indirectly through their indispensable contributions to the success of our best overall theory of the world. Nevertheless, an aura of mystery remains: the justification for such claims seems to be different in kind and much less direct than the justifications we have for claims about goings-on in the actual world.

The best way to render putative synthetic a priori truths unmysterious is to naturalize them. This requires identifying them within the scientific worldview, rather than merely showing that they are necessary to underpin that worldview. I shall offer a naturalistic treatment of recombination principles in the context of EQM. According to this proposal, the *unitary* evolution described by the Schrödinger equation is best understood as more akin to a recombination principle than to a law of any individual world. The Schrödinger equation says how things are across physical modal space, rather than just saying how things are at the actual world.

Lewis appealed to his principle of recombination in order to capture what he called the requirement of plenitude: a modal realist should ensure that there are 'worlds enough, and no gaps in logical space' (Lewis 1986b). The Schrödinger equation plays the analogous role in EQM. For Everettians, what the quantum-mechanical principle of unitarity ensures is that there is an Everett world for every dynamically allowed history. It thereby rules out multiverses with worlds excised via any qualitative criterion. For example, it rules out multiverses including all the worlds except those containing both cats and comets. This is strikingly reminiscent of Lewis's requirement of plenitude. Still, the theoretical roles of unitarity and the Lewisian recombination are not identical: whereas the Lewisian recombination principle applies to fundamental entities, thanks to the emergent nature of contingency in quantum modal realism, the plenitude of possibilities guaranteed by the Schrödinger equation is at the derivative level.

What recombination principles do hold in quantum modal realism? The Schrödinger equation does not have the explicit form of a principle of recombination as characterized earlier: it is not a material conditional of the right kind. But the Schrödinger equation can nonetheless help to explain the holding of recombination principles of that form. Unitarity makes sure that any dynamical process will produce sets of worlds for each of the possible outcomes of that process. In cases where there is an appropriate symmetry over these possible outcomes, the different outcomes that are obtained by reflecting the subsystem along the appropriate axis of symmetry will correspond to particular instances of free recombinability.

My proposal, to be specific, is that the Schrödinger equation, combined with environment-induced decoherence, helps explain the truth of a variety of situation-specific macro-level recombination principles. Decoherence approximately privileges the position basis (see section 2.3 for more details), so the resulting recombination principles will be like Lewis's, involving recombination of spatial regions. Macroscopic objects have highly decohered wavefunctions and accordingly have freely recombinable spatial positions. The recombination principle to be derived will be a material conditional connecting (inter alia) facts about the spatial properties of specific kinds of decohered systems. We may need to add additional contextual factors into the explanation: for example, whatever facts determine that decoherence produces states highly localized in position.

An example may help to illuminate matters. Consider a box containing a single electron (Figure 1.1, top). In classical physics, the particle may stay on one side of the box indefinitely. But in quantum physics, the unitary evolution spreads the wavefunction out to fill the box (Figure 1.1, bottom).

Measurements decohere, and hence localize, the electron: all points in the box are possible outcomes of measurements on the electron position. Accordingly, in EQM all points in the box are physically possible locations for the particle. Generalizing to the multi-particle case, we obtain the result that all combinations of particle locations are physically possible. This is the sort of recombination principle we were after. We can also see in outline how to generalize such principles to more complex macroscopic configurations. For every possible game of chess involving fewer than a hundred moves, there is a pair of people in some world who play that particular combination of moves. For every possible temporal sequence of outcomes on a roulette wheel, there is some world with exactly that sequence of outcomes.

Recombination principles partly grounded in unitary evolution and in decoherence are approximate, miscellaneous, and non-fundamental. They are approximate, because if a system's wavefunction is insufficiently decohered, then recombination does not hold true of the system (even approximately). They are miscellaneous, because different variables may be recombined in different patterns depending on the physical structure of

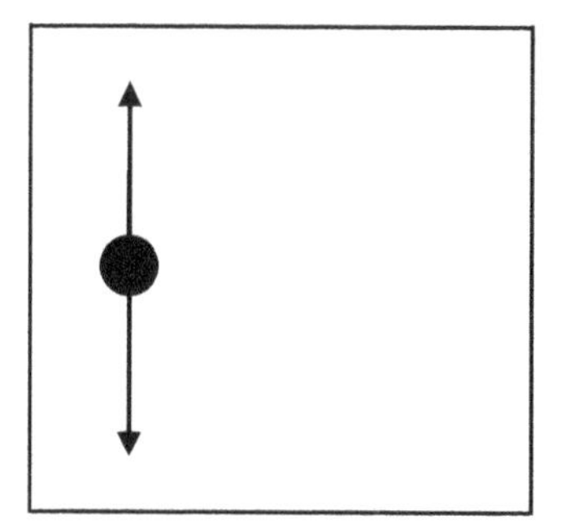

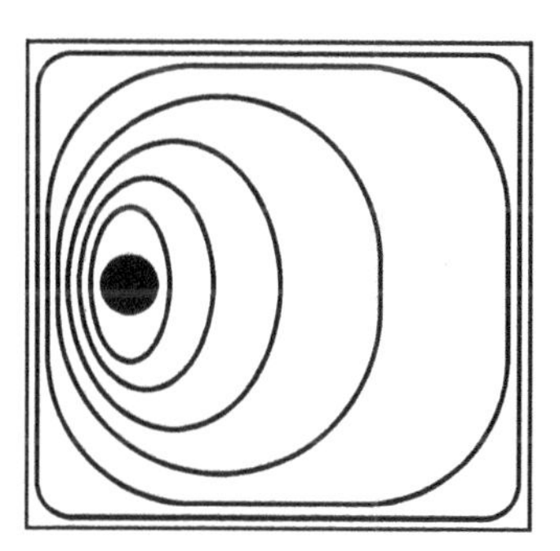

Figure 1.1. Box containing classical particle (top) vs quantum particle (bottom). Not to scale.

the particular system modelled. And they are non-fundamental, since these principles are derived rather than axiomatic, and since they concern a non-fundamental ontology of emergent Everett worlds.

What we do not have in the Everettian context is free recombination of the fundamental properties of the theory. There is no free recombination within the universal state itself, only the evolution determined by the Schrödinger equation. This means that the quantum modal realist picture fails to include any properties that respect a free-recombination-of-the-fundamental principle. It is frequently assumed that free recombinability is a central part of the theoretical role of the fundamental, although Wang (2016) surveys the available arguments and sounds a note of caution. But quantum modal realists may simply deny that this is a genuine aspect of the fundamentality role, perhaps taking a lead instead from physics in their use of the term 'fundamental'. In this denial they may take heart from the technical difficulties affecting naïve unrestricted recombinatorial reasoning: for a striking recent presentation of these difficulties, see Fritz (2017).

To sum up: quantum modal realism incorporates a radical and powerful modal epistemology wholly unlike that of any other available theory of modality. For quantum modal realists, our knowledge of metaphysical possibility is no more and no less mysterious than our knowledge of chemistry. It needs no distinctive sources of evidence and no distinctive methodology. In this sense, quantum modal realist epistemology is modal epistemology naturalized. Exactly how much revision of our pre-theoretic modal opinions is required by quantum modal realism is an empirical matter, but there is no reason to think that this revision makes any difference to our core modal beliefs, the ones that we care about and act upon.

1.9 Actuality

Possibly the single most distinctive feature of modal realism is its indexical analysis of actuality. The actual world is not intrinsically metaphysically privileged according to modal realists: it is simply the Everett world that you and I happen to inhabit. Everettians recognize no sense in which any Everett world is physically or metaphysically privileged over any other, so correspondingly in quantum modal realism there is no sense in which any metaphysically possible world is privileged over any other. In this respect, quantum modal realism directly mirrors Lewisian modal realism; however, both theories differ radically from more orthodox approaches to actuality that have it extensionally equivalent to reality. Prominent examples of this orthodoxy are Plantinga (1974) and R. M. Adams (1974).

I will not reprise here Lewis's persuasive defence of the legitimacy of the real/actual distinction against complaints that 'everything real is actual' is analytically true, or his fallback proposal to use 'thisworldly' as an alternative technical term when formulating his theory (Lewis 1986b: 97–101). Quantum modal realists, like Lewisian modal realists, can give straightforward explanations of why we might intuitively have thought that reality and actuality coincided. Actuality as countenanced by quantum modal realists closely resembles reality as a whole as countenanced by traditional actualist accounts of modality. Until EQM provided us with compelling scientific grounds to believe in a reality that contains much more than that traditional vision of the contents of actuality, we had no reason to think that reality and actuality diverged. But in the Everettian picture, reality extends far beyond traditional visions of the contents of actuality, and this casts the equation between reality and actuality into serious doubt. A real–actual distinction allows us to vindicate our view of what is actual even as we give up our previous view of what is real.

As well as vindicating in general terms our judgements about the extension of actuality, the real–actual distinction vindicates a broad range of other kinds of judgements that seem threatened by the quantum modal realist picture. The pattern is the same as before: we recapture some familiar range of judgements by regarding them as judgements about actuality, although they are false if interpreted as judgements about reality. This section will explore one case study of central interest: the challenge that quantum modal realists face in accommodating the kinds of prudential and ethical values which we ordinarily take to guide our actions.

A distinctive line of objection to Lewisian modal realism has targeted its ethical implications. If the totality of what exists will be the same no matter how we act, how can it matter what we do? A very similar line of objection may be directed at quantum modal realism. In the remainder of this section, I shall respond to this objection by arguing that the metaphysics of ethics is not a topic which can be addressed in isolation from the metaphysics of modality. Instead, in accordance with the confirmational holist methodology outlined in section 0.4 of the introduction, we ought to allow defenders of revisionary modal-metaphysical theories to develop ethical theories that complement their metaphysics and then assess these theories alongside the metaphysics, as a package deal. The package deal of quantum modal realism with what I shall call an actualized conception of value has the same 'first-order' ethical consequences as a combination of a one-world account of modality with a non-actualized conception of value.

The ethical objection to modal realism was first raised by Robert M. Adams, as follows:

> [O]ur very strong disapproval of the deliberate actualizing of evils... reflects a belief in the absolutely, and not just relatively, special status of the actual as such. Indeed, if we ask, 'What is wrong with actualizing evils, since they will occur in some other possible world anyway if they don't occur in this one?', I doubt that the indexical theory can provide an answer which will be completely satisfying ethically.
>
> R. M. Adams (1974: 215–16)

From this passage we can extract the following normative argument against modal realism:

1) If the totality of existence will be the same whatever I do, it doesn't matter what I do.
2) Given modal realism, the totality of existence is non-contingent.
3) Given modal realism, it doesn't matter what I do.

The normative argument comes in a number of flavours. It can be given a moral flavour: why should I do the right thing, if the totality of existence will be the same whatever I do? It can be given a prudential flavour: why should I do the sensible thing, if the totality of existence will be the same whatever I do? It can even be given a fatalistic flavour: why should I continue to live, if the totality of existence will be the same whether I live or die?

Lewis responded to the normative objection by appealing to irreducibly egocentric desires, in which an agent desires that *they themselves* should have some particular property:

> Wanting to live is not wanting that a kind of thing happen, somewhere in the worlds and never mind where; it is an egocentric want, a want that I myself should have a certain property. The appropriate way to give the content of my desire is not by a condition that I want the entire system of worlds to satisfy, but by a condition that I want myself to satisfy.
>
> Lewis (1986b: 125)

I think this answer is satisfying for the prudential and fatalistic versions of the normative argument, but it seems to founder when applied to the ethical objection.[14] Doing the ethically right thing, at least on orthodox views of the matter, cannot be captured in terms of satisfying any egocentric desire.[15] Accordingly, I shall set aside the specific analysis of desire as egocentric and

[14] A particularly virulent version of the normative argument is the causal version deployed against Lewisian modal realism by Heller (2003): by causing good in your own world, Heller argues, you cause equal and opposite harm elsewhere. Depending on your favoured moral theory, this might entail that no action can be good, or that all actions are equally right/wrong. To defuse this specific problem, quantum modal realists should deny the causal claim, for example by appealing to a counterfactual theory of causation and an Everettian counterfactual semantics (see sections 1.6 and 1.3 respectively).

[15] It is no surprise that this version of the argument is the one on which recent exchanges concerning Lewisian modal realism have focused; see for example Heller (2003) and Cresswell (2005).

suggest a more general response: if it works, then it applies to all versions of the objection.

Consider what happens if we replace 'existence' with 'actual existence' in the normative argument. The argument then lapses—the totality of actual existence is contingent, since other worlds are possible, so while premise 1* is true, premise 2* is false:

1*. If the totality of actual existence will be the same whatever I do, it doesn't matter what I do.
2*. Given modal realism, the totality of actual existence is non-contingent.

We can then deny the original unactualized premise 1, and explain away its plausibility as deriving from our running it together with the true premise 1*. Both arguments are thereby rendered unsound, and the normative argument is defeated.

Is it really plausible, though, to deny the unactualized premise 1? It can be derived, after all, from the following plausible-looking version of consequentialism, which corresponds to what Jack Smart called a 'truly universalistic ethics' (Smart 1984):

Existence-Consequentialism (EC): The right act to perform is the act with the best consequences for the totality of existence.

The response from the quantum modal realist is to deny EC and replace it with an appropriate actualized formulation of consequentialism:

Actuality-Consequentialism (AC): The right act to perform is the act with the best consequences for the actual world.

Quantum modal realists will deny EC and explain away its plausibility as deriving from our running it together with the plausible AC, as they did for 1 and 1* respectively. In a consequentialist setting, the combination of modal realism with AC has the same first-order normative implications as the combination of modal actualism with EC. So consequentialist normative beliefs do not, by themselves, distinguish between modal realism and modal actualism.

Can modal realism also be defended from the ethical objection in the context of non-consequentialist ethical theories? Lewis thought non-consequentialists had no case to answer in the first place: 'if modal realism makes a problem for anyone, it is for utilitarians' (Lewis 1986b: 127). But this does not seem quite right. I will discuss two cases—virtue ethics, an easy case, and deontological ethics, a harder case.

Modal realists who are also virtue ethicists do look to have a ready answer to the normative argument. Whether or not modal realism is true, our own virtue is actual virtue. This could be resisted: perhaps modal realism undermines the very possibility of virtue. But making this case would require rejecting the modal realist's analysis of modality on independent grounds, so I agree with Lewis that modal realism makes no special trouble for virtue ethics.

Lewis does not address deontological ethical theories, but there are prima facie concerns about their compatibility with modal realism. For example: can we ever coherently wish every agent to act on a certain maxim, if at the fundamental level it is non-contingent who does and who does not act on that maxim? Such concerns can be assuaged by the same strategy as was applied to consequentialism. We distinguish between two forms of a toy deontological theory:

Existence-Deontology (ED): Act only on the maxims that you can coherently wish would be acted on by all existing agents.
Actuality-Deontology (AD): Act only on the maxims that you can coherently wish would be acted on by all actual agents.

As before: AD taken together with modal realism has the same first-order ethical consequences as ED taken together with modal actualism.

Mark Heller is unimpressed with the sorts of manoeuvres I have suggested, since they conflict with some theoretical principles that he maintains are highly intuitively plausible: 'As long as the people in other worlds are real people they must figure in our moral calculations, regardless of whether they are our worldmates' (Heller 2003: 20–1). 'I see no plausibility whatsoever to the idea that mere spatial proximity should have any moral significance, and, likewise, I see no plausibility whatsoever to the idea that the worldmate relation itself should have any moral significance' (ibid.: 4). Faced with this sort of insistence that the non-actualized versions of consequentialism or deontology are correct and the actualized versions are incorrect, quantum modal realists may respond by adopting a divide-and-conquer strategy. Intuition, they may say, is reliable with regard to first-order normative beliefs; but quantum modal realism, via an actualized ethics, can capture such beliefs. Intuition, on the other hand, is not reliable with regard to higher-order normative beliefs including the theoretical principles adduced by Heller. Any intuitive support for a universalistic ethics over an actualized ethics is accordingly very weak.

I hope the general shape of the actualization strategy is clear. It can evidently be applied beyond normative concerns, to any philosophical context where the greatly enlarged reality envisaged by quantum modal realists appears to conflict with intuitive or widely held theoretical principles. Quantum modal realists can capture whatever explanatory power such principles have by appealing to their actualized versions, while denying the probative force of intuition with respect to the original unactualized principles. This manoeuvre, if it needs any special justification at all, may be justified by appeal to the uniquely central role of the actual world for our interests: we can neither causally influence nor be causally influenced by events in other Everett worlds, real though those events are.

This concludes my presentation of the core features of quantum modal realism. In chapter 2, I turn to the physical theory on which quantum modal realism relies: EQM.

2
Diverging Everettian Quantum Mechanics

2.1 Introduction

This chapter sets out the way of understanding quantum theory—*diverging Everettian quantum mechanics*—which is incorporated into quantum modal realism. Quantum theory, as with other theories in fundamental physics, does not wear its interpretation on its sleeve. Even amongst approaches to quantum theory in the broadly Everettian no-collapse tradition, there are numerous important differences of detail. Before we can fully evaluate the bearing of EQM on controversial metaphysical questions, we will need to settle on a consistent and plausible version of the theory. This is easier said than done; the details of the ontological picture underlying EQM have proved highly controversial, even amongst its defenders.

After a sketch of the history and the basic physics of the approach in sections 2.2 and 2.3, I argue in section 2.4 that the metaphysics of EQM is underdetermined by the physics: the perspicuous *decoherent histories* formalism permits both an interpretation according to which Everett worlds do not literally overlap and an interpretation according to which they do. In my view, Everettians have compelling reason to adopt a non-overlapping—or *diverging*—version of EQM. Sections 2.5 and 2.6 explain why. Section 2.7 summarizes the arguments of the chapter.

2.2 Background

The physical theory drawn on by quantum modal realists is Everettian quantum mechanics of the decoherence-based variety, as recently championed by the 'Oxford Everettians': Simon Saunders, David Deutsch, David

The Nature of Contingency: Quantum Physics as Modal Realism. Alastair Wilson, Oxford University Press (2020).

DOI: 10.1093/oso/9780198846215.001.0001

Wallace, and Hilary Greaves. In this section I give a quick sketch of the history and conceptual basis of modern decoherence-based EQM.

EQM has some roots in early work of Erwin Schrödinger on the interpretation of his newly developed *wavefunction* approach to quantum theory, but the key idea was first unambiguously presented by Hugh Everett III in his Princeton doctoral thesis (Everett 1957a). By the time Everett was writing, quantum mechanics was already overwhelmingly empirically successful: it had been used to explain a vast range of properties of atomic matter, and had led directly to the development of nuclear power and weaponry. Yet it was still dogged by conceptual difficulties, chief amongst them the notorious *quantum measurement problem.*

Here is a quick sketch of the measurement problem. Although transition probabilities calculated using the mathematical apparatus of quantum mechanics provide astonishingly accurate matches to the statistical results of experiments, the procedure actually used by physicists when applying the formalism (call it the *quantum algorithm*) makes essential use of the operational concept of a measurement. Typically, the formalism provides a specification of the *quantum state* of a target system (as a vector in a Hilbert space, up to normalization and phase); such states may be analysed into *superpositions*—vector sums—of distinct components, along with *amplitudes* for each component. The *Born rule* tells us that, when a measurement is made, the probability of a given result is equal to the squared-amplitude of the component of the state that corresponds to that result, according to some appropriate decomposition.[1] The quantum algorithm, of course, allows for repeated measurements on a system. But in calculating probabilities for outcomes of a second measurement, the algorithm tells us not to reuse the initial state of the system prior to the first measurement. Rather, it instructs us to take the initial state A, evolve A forward in time to state B, pick whichever component C of B corresponds to the result actually obtained, renormalize C and evolve it forward in time to state D, and then calculate probabilities for outcomes of the second measurement according to the amplitudes of the various components of D. Components of B corresponding to results not obtained in the first measurement are discarded; the algorithm makes no further use of them.

The way in which the concept of measurement appears in the quantum algorithm is *prima facie* deeply puzzling. As scientific realists, we look to our

[1] The vagueness of this phrase points to the 'preferred basis problem', described in more detail below.

best fundamental physics to tell us what reality is like; but in the quantum algorithm we find only an 'ad-hoc, undermotivated, inherently approximate' (Wallace 2010a: 22) operational procedure for predicting results of measurements. What we would like is a picture of reality that explains why using the quantum algorithm leads us so reliably to correct predictions. Insofar as the algorithm's effectiveness has been experimentally confirmed, but we still lack any plausible picture of reality capable of underwriting it and explaining its success, we face the quantum measurement problem.

Perhaps the most obvious way to explain why the quantum algorithm works so well is to imagine that the stage of the quantum algorithm where components not corresponding to the actually measured result are discarded tracks some genuine physical process. That requires that the measurement process should destroy the components corresponding to the not-actually-measured outcomes. The suggestion is that some kind of *collapse* occurs when a measurement is made; some real physical process, it is argued, must pick out the actual result from the merely possible results. However, such a collapse process raises problems of its own. There appear to be countless distinct but empirically indistinguishable ways that such a process could work—which in itself might give us grounds for suspicion—but in any case, all of them face well known and apparently intractable compatibility problems with the principle of Lorentz covariance enshrined in special relativity.[2]

An enormous number of solutions to the quantum measurement problem have been attempted, but most have turned out to be incoherent, to raise intractable conceptual problems of their own, to sacrifice the principle that reality is mind-independent, and/or to simply fail to solve the problem. Those theories which do resolve the measurement problem while remaining coherent—the first and still the most well known of which is *pilot-wave theory*, or *Bohmian mechanics*, originally proposed by Louis de Broglie (1927) and given canonical form by David Bohm (1952a, 1952b)—typically do so by adding some additional dynamical structure with no obvious correlate in the quantum formalism. Such theories suffer from the same problems with Lorentz covariance as do interpretations involving objective collapse of the wavefunction, as well as being liable to charges of arbitrariness in the choice of extra structure.

[2] Maudlin (1994) rigorously summarizes these difficulties; however, see Myrvold (2002) for a defence of compatibility between special relativity and the broader collapse approach. The 'flash' variant of the Ghirardi-Rimini-Weber theory (Tumulka 2006) avoids the clash with relativity.

The dominant figure in the interpretation of quantum mechanics, Niels Bohr, rejected all requests for an explanation of the success of the quantum algorithm as conceptually confused, thereby undercutting both the approach that seeks to provide a dynamical explanation of collapse and the approach that supplements the quantum description with additional variables. The 'Copenhagen interpretation' preached by Bohr and Heisenberg painted the measurement problem as a pseudo-problem. When we give up our prejudices derived from classical physics and classical metaphysics, he urged, we will see that no quantum measurement problem remains. Bohr promoted a conceptual framework of 'complementarity' that was at best unclear and at worst actively obfuscatory, and he aggressively suppressed alternative proposals; the history is told engagingly, though in different ways, by Cushing (1994) and by Beller (1999).

Discontented with the Bohrian party-line understanding of quantum mechanics, and discontented in particular with the problems concerning wavefunction collapse that Bohr refused to acknowledge, Everett came up with a remarkable idea whilst a PhD student at Princeton in the mid-1950s. What if quantum states never really collapse, and the appearance of collapse is a purely perspectival phenomenon? The quantum dynamics generically evolves quantum states into *superpositions*; where the orthodox interpretation took superposed quantum states to represent single systems with unfamiliar indeterminate properties, Everett proposed taking superposed states to represent multiple systems each with familiar determinate properties. In other words, the central idea of EQM is to replace indeterminacy with multiplicity.

Everett himself did not use the term 'many worlds' to refer to his own theory.[3] He preferred to gloss the approach as the 'Relative State Interpretation': he understood it not as multiplying individual systems themselves but as multiplying relational states of individual systems, so that an electron could be spin-up relative to one substate and spin-down relative to another, and a cat could be dead relative to one substate and alive relative to another. Subsequent Everettians have often balked, though, at this unfamiliar relational metaphysics. Instead, EQM first came to prominence when it was championed by Bryce DeWitt under the guise of the 'many worlds interpretation' in DeWitt (1968, 1970); early work in the area is collected in an anthology edited by DeWitt and Graham (1973) titled 'The Many

[3] Conroy (2012) provides a careful review of Everett's preferred ways of describing his own theory.

Worlds Interpretation of Quantum Mechanics'. In line with the majority of contemporary Everettians, and without further argument, I will work with the many-worlds formulation of the approach.[4]

Everett's own presentation of EQM left many important questions unanswered; I shall mention two in particular. Everett had provided a proof that the amplitudes of branches where the frequencies of observed outcomes diverge from the quantum amplitudes themselves tend towards zero in the infinite limit; however, this theorem is generally considered inadequate since the link it forges between frequency and amplitude itself holds only with high amplitude. Moreover, there is a powerful intuition that probabilities in a multiverse should be linked to relative number of branches, which would conflict with the Born rule. At the very least, more needs to be said in order to provide a fully explanatory account of probability in EQM. In addition, Everett's formulation of the theory required the explicit stipulation of a preferred basis: essentially, a set of physically privileged quantities by means of which the individual branches of the quantum state are distinguished. This preferred basis seems to be unwarranted by the quantum formalism, to run contrary to the spirit of the theory, and to undermine EQM's claim to Lorentz covariance.

Deutsch (1985) endorsed Everett's stipulative solution to the preferred basis problem, describing this solution as the appeal to an 'interpretation basis', and he added the suggestion that the problems with probability could be resolved by adding an additional continuous infinity of worlds into every quantum interaction. For each branch of the quantum state, there are infinitely many universes, with the relative numerosity of universes having different qualitative characters fixed by the Born rule measure associated with those qualitative characters. Now we have infinitely many worlds corresponding to each single physically possible history, but some of these infinities are larger than others. Here is the full proposal.

> I propose a slight change in the Everett interpretation:
>
> *Axiom 8.* The world consists of a continuously infinite-measured set of universes.
>
> By a 'measured set' I mean a set together with a measure on that set. The interpretation of the state... will be that the set of universes consists of n_1

[4] See Conroy (2018) on the prospects of adapting quantum modal realism to the relative-state approach.

> disjoint subsets, where the a_1th subset is of measure $|C_{a_1}|^2$. Each of these subsets, which I shall call a *branch*, consists of a continuous infinity of identical universes. During the model measurement, the world has initially only one branch, and is partitioned into n_1 branches. The branches play the same role as individual universes do in Everett's original version, but the probabilistic interpretation is now truly 'built in'. Deutsch (1985: 20)

Deutsch's proposal attracted plenty of criticism, primarily because of the arbitrariness of the interpretation basis (and the conflict with special relativity that it seems to introduce) and because of the incorporation of an additional proliferation of universes with no correlate in the quantum-mechanical formalism. The explicit introduction of a continuous infinity of diverging universes in particular seemed to run counter to the spirit of EQM, as expressed in DeWitt's often-quoted remark that that 'the mathematical formalism of the quantum theory is capable of yielding its own interpretation' (DeWitt 1970). It is not hard to see why Deutsch's theory was unpopular; if we were eventually to decide that quantum theory has to be modified by adding new structure at the fundamental level, we would be inclined to prefer a way of modifying it that does not involve introducing infinitely many additional distinct universes.

In a similarly inflationary vein, Albert and Loewer (1988) described a 'many minds' version of EQM which, like Deutsch's proposal, violated Lorentz covariance and introduced unwelcome new dynamical structure (in this case a set of stochastically evolving minds), and which in addition incorporated an explicit mental–physical dualism. This package of views is very remote from Everett's original vision, and it is unsurprising that Albert and Loewer's theory has attracted few advocates as an alternative to unitary quantum mechanics; Albert and Loewer never themselves endorsed it. EQM is in at least one literal sense a theory of many minds, as it entails the existence of many more people than we previously suspected and each of these additional people comes equipped with a mind. But using the term 'many-minds' to characterize EQM, as Lockwood (1989) and Papineau (1996) also do, is misleading: mentality need not enter at any stage into the formulation of the theory itself.

EQM suffered for a time from the perception, encouraged as we have seen by some of its advocates, that additional structure (whether of worlds or minds) has to be added to the theory to make it coherent. The most significant step towards a plausible version of EQM came when, in the early 1990s, progress in technical work on *decoherence* was applied to the preferred basis problem in EQM by Saunders (1993, 1994, 1995). Decoherence theory can be used to model the quantum-mechanical interactions between a system and its environment; Zurek (2002) gives an accessible introduction. The essence of decoherence is that a broad range of quantum systems evolve in such a way as to suppress to a negligible level the interference terms representing interactions between components of the state of the system corresponding to distinct macroscopic properties. This suppression of interference terms happens extremely quickly; simple models have it occurring at the order of 10^{20}s (Zurek 2002).

The most frequently studied cases of decoherence arise from systems of interest becoming entangled with their environments through interaction with them; it can be shown in numerous different model scenarios[5] that this entanglement leads to suppression of interference between macroscopically distinct states of the system of interest. However, the decoherence-based solution to the preferred basis problem does not essentially involve appeal to the distinction between a system and its environment. Conditions for the effective suppression of interference effects can be stated even without any system/environment distinction, as is done in the *decoherent histories* formalism for quantum theory. Since this formalism lends itself particularly well to discussion of conceptual issues in EQM, it is briefly reviewed in the next section and employed in the subsequent discussions of probability.

Although decoherence suppresses interference between macroscopic superpositions, it does not eliminate this interference altogether. The idea behind decoherence-based EQM is that a preferred basis is *approximately* picked out by decoherence, to a degree of approximation easily high enough to explain the fact that superpositions of macroscopic states are unobserved and effectively unobservable. If we consider some individual component (in the decoherence basis) of some well-decohered state, and assess probabilities for the forwards time-evolution of that component, then the other components of the state have a negligible influence; it makes no significant difference to our calculated transition probabilities whether or not we consider these other components. The approximate nature of decoherence gives rise to one of the most distinctive features of contemporary EQM. This is that Everett worlds, as represented by decoherent histories, are to be understood as *non-fundamental*. The universal quantum state is the fundamental entity

[5] Crull (forthcoming) surveys some important categories of decoherence models.

of EQM, and individual Everett worlds are *emergent*[6] or *derivative* entities, grounded in the universal state.

Initial formulations of EQM with a preferred basis, such as Deutsch's 1985 proposal, treated Everett worlds as fundamental. By contrast, the Oxford Everettians have come to embrace a conception of Everett worlds as emergent. In what Saunders (2010a) has called the 'killer observation', Wallace (2003a, 2010a, 2012) assimilated the status of Everett worlds to that of emergent ontology more broadly, arguing that approximateness is endemic in higher-level ontology. Nobody expects the entities of chemistry, biology, and astrophysics to have precise definitions in terms of fundamental physics: it is widely recognized that there is unavoidable indeterminacy as to whether a particular hydrogen atom is part of a virus, as to when one species evolves into another, and as to where the boundaries of a nebula lie. The same may holds for Everett worlds.

Wallace (2012) gives an elegant statement and extended defence of a decoherence-only version of EQM that treats Everett worlds as emergent entities. Although it remains controversial in the interpretive literature whether decoherence is by itself sufficient to resolve the preferred basis problem, there is a growing consensus that the success of EQM depends on it; many-worlds or many-minds theories that posit additional fundamental structure would not be worth the theoretical costs. Accordingly, in this book I work exclusively in the context of a decoherence-only version of EQM, making extensive use in the quantum modal realist system of the contrast between the single fundamental quantum state and the multiplicity of non-fundamental Everett worlds which emerge from it.

I aim to remain as neutral as possible on the precise way we should think about the ontology of the quantum state, and about its relation to spacetime. In our current state of ignorance of Planck-level physics, we still have no clear idea of how—or even whether—spacetime itself should be given a quantum description. Accordingly there is no consensus on whether the quantum state is best thought of as an entity in its own right, or as the state of something else—perhaps of some fields on spacetime or on some less familiar space. The positive ontology associated with EQM is obviously of critical importance for a complete formulation of the theory, but it will

[6] Some conceptions of emergence involve more than being non-fundamental yet of explanatory value, for example by requiring in addition a failure of supervenience of whole on parts or novel fundamental laws applying to higher-level entities. These 'strong' forms of emergence are not what Wallace has in mind and I set them aside here.

prove relatively unimportant for the applications of the theory to the explanation of modality pursued in this book. Nonetheless, it is worth briefly describing what I see as the two leading proposals for the ontology of EQM.

The ontological picture most often combined with EQM is *wavefunction realism*. According to this approach, the quantum state is a new kind of fundamental entity: a complex-valued field on the *configuration space* of a system. For the simplest case of an N-particle system, the configuration space would be 3N-dimensional, encoding all of the positions of all of the particles into a single point of the space. The quantum state 'lives' on the space, evolving over time according to the Schrödinger equation and fixing the probabilities for locations of each of the particles at any given time. A notable variant of the view involves 'multi-fields' on ordinary physical space, which assigns field values to regions rather than to individual points. See Chen (forthcoming) for a review of the state of the art in wavefunction realism.

The primary alternative to these wavefunction realist approaches is to focus attention on physical spacetime, and to regard the quantum state as the state of spacetime itself. Quantum theory specifies properties of spacetime, rather than properties of a field on space that evolves over time as in wavefunction realism. An important defence of this *spacetime state realism* is given by Wallace and Timpson (2010); see also Wallace (2012) for additional discussion. Spacetime state realism involves associating a *density operator* with each spacetime region, representing the state of that region. Entanglement in this picture has the generic consequence that the states of regions are *non-separable*: the state of a region does not supervene on the states of its non-overlapping subregions. This sets spacetime state realism apart from wavefunction realism, where separability is maintained.

There is little point, I think, in seeking a fundamental ontology for quantum physics in the absence of a well-developed theory of quantum gravity. According to some approaches to quantum gravity, spacetime itself is non-fundamental; if such approaches are vindicated, spacetime state realism will obviously stand in need of some modification. The exact content of wavefunction realism likewise depends sensitively on what sorts of physical entities—particles, fields, or something different again—are posited by successors to the standard model. Different fundamental physical theories give rise to very different configuration spaces. However, of all interpretations of quantum theory EQM is the best placed to flexibly accommodate unexpected future developments. It is tied neither to particles, or fields, nor even to ordinary space and time. It relies only on the core features common to any quantum theory—the unitary evolution encoded in the Schrödinger equation, and the link between probabilities and amplitudes encoded in the Born rule. Accordingly I think quantum modal realism may remain almost entirely neutral on the underlying ontology of the fundamental quantum state. All the action takes place at the level of the non-fundamental Everett worlds. It is time to turn to the decoherent histories framework, our best tool for modelling these Everett worlds.

2.3 Decoherent Histories

The decoherent histories formalism is an elegant and perspicuous formulation of no-collapse versions of quantum mechanics. It is widely favoured by Everettians, since it enables reasoning about classes of entire histories. This section, which draws on A. Wilson (2012), comprises a simple account of the decoherent histories formalism. The aim is to frame questions about the representational significance of particular terms in the equations using just enough technicality to connect my discussion up with the existing literature on decoherent histories.

Most elementary formulations of quantum mechanics are given in the *Schrödinger picture*: a wavefunction evolves over time, and operators (representing types of measurement) remain constant. The decoherent histories formalism makes use instead of the intertranslatable *Heisenberg picture*: the quantum state is constant, while operators (associated with observable quantities) change over time. In order to characterize histories, we consider observables corresponding to the whole state of a world at a time. We represent the alternative possibilities by orthogonal projection operators $\widehat{P}_\alpha$, summing to unity:

Orthogonality: $\widehat{P}_\alpha \widehat{P}_\beta = \delta_{\alpha\beta} \widehat{P}_\alpha$
Unitarity: $\sum_\alpha \widehat{P}_\alpha = 1$

We can think of the first of these conditions as ensuring that the possibilities represented are mutually exclusive, and the latter as ensuring that they are exhaustive. If a partition into projection operators meets these conditions, it is called a *coarse-graining*. Coarse-grainings generate sets of *histories*, time-ordered sequences of time-dependent projection operators:

Histories (defn.): $C_{\underline{\alpha}} = \widehat{P}_{\alpha_n}(t_n)\, \widehat{P}_{\alpha_{n-1}}(t_{n-1}) \ldots \widehat{P}_{\alpha_0}(t_0)$

According to contemporary decoherence-based EQM, as propounded by Saunders, Wallace and others, these histories are the kind of thing that can in principle represent individual worlds, and sets of histories are the kind of things that can in principle represent complete multiverses.

Sets of histories must meet further conditions if they are to accurately represent a set of causally isolated Everett worlds. In particular, the worlds must be sufficiently well decohered, sufficiently dynamically independent of one another. For our purposes this requirement is best captured by the *medium decoherence condition* (Gell-Mann & Hartle 1993).[7] A set C obeys medium decoherence if and only if for any histories $C_{\underline{\alpha}}$, $C_{\underline{\alpha}'}$ in C (where ρ is a density operator representing the initial quantum state of the universe, and Tr is the matrix trace operation):

Medium Decoherence: $\mathrm{Tr}(C_{\underline{\alpha}}\rho\, C^{\dagger}_{\underline{\alpha}'}) \approx \mathrm{Tr}(C_{\underline{\alpha}}\rho\, C^{\dagger}_{\underline{\alpha}})\delta_{\underline{\alpha}\underline{\alpha}'}$

Sets of histories meeting this condition are called decoherent history spaces; Everettians use them to represent multiverses such that the individual worlds in the multiverse are effectively causally isolated from one another. Individual histories in these sets are called decoherent histories; Everettians use them to represent universes like our own.

Distinct histories in a decoherent history space will agree up to some time, but differ thereafter: this process of differentiation is known as *branching*. The next section will consider the metaphysics of branching in detail, but some initial observations may help to situate the reader for whom the idea is unfamiliar. Branching occurs whenever decoherence becomes sufficient to render different histories effectively causally isolated, for example when a dust particle becomes entangled with a radiation bath environment so that the components of the particle's state corresponding to superposition of macroscopic properties become negligible compared to the components corresponding to reasonably precise macroscopic properties. Branching may be thought of as a transition from a particle not yet correlated with its environment and with a relatively indeterminate location, to multiple particles correlated with their environments, each with a relatively determinate location. Interaction, either internally to a system or with an external environment, is required for decoherence to take place: an isolated particle (in non-relativistic particle mechanics) will not decohere. In general terms,

[7] A number of different decoherence conditions have been formulated in the literature. For discussion in the context of EQM, see Wallace (2012: ch. 3).

the more complex the interactions a system undergoes, and the more sensitive the dependence of the evolution of a system on its initial state, the more rapidly that system will decohere.

The emphasis placed in decoherence-based EQM on the effective causal isolation of distinct Everett worlds may seem to sit oddly with some familiar glosses on EQM. Popular discussions of the theory often place great weight on its giving a causal explanation of quantum interference effects. The suggestion is typically that the other worlds causally interact with the actual world, causally contributing to the events that we observe. I think this suggestion is mistaken. We ought not to think of a non-contingent subject matter in causal terms, and accordingly the explanations that EQM affords of quantum interference phenomena are better thought of as non-causal explanations than as causal explanations. Even though we do not causally interact with other branches, their existence is part of the best overall explanatory story behind the evidence that we have. This non-causal perspective on quantum interference does not detract from the key explanatory role that decoherence plays in EQM in generating an emergent structure of worlds, but it does cast that explanation in a different light, as one which cites features of a law-governed pattern rather than features of a causal process.

In an attempt to better understand these non-causal explanations, we might try to generalize the counterfactual theory of explanation to allow for explanatory connections which stand to counterpossible counterfactuals just as causal connections stand to non-counterpossible counterfactuals. This project is explored in Reutlinger (2017), Schaffer (2016) and A. Wilson (2018b). But it is certainly not compulsory for quantum modal realists: in the recent literature there are a number of accounts of non-causal explanation that do not rely on any specific connection with counterfactuals. These include the unificationist explanations of Friedman (1974) and Kitcher (1981), the structural explanations of McMullin (1978), and the grounding explanations of Fine (2012b). In principle, we could seek to model the non-causal explanation of the features of one Everett world in terms of the features of other Everett worlds by drawing on any of these broader accounts of non-causal explanation.

The non-causal understanding of quantum interference that I have advocated is an unfamiliar one. It involves a substantial adjustment to our conception of how physics explains contingent events. Some aspects of actual events (including, for example, their values of conserved quantities) are best explained in terms of their actual causes, but other aspects of actual events (including, in particular, their chances) are best explained in terms of their place in a larger overall mosaic of possible events. In this respect my suggested picture is more revisionary than one-world approaches to quantum mechanics (even those which introduce primitive stochasticity) and more revisionary than many-world approaches which locate all of the Everett worlds within a single causal web. Still, I think the non-causal understanding of interference is uniquely well-suited to the quantum modal realist's vision of the relation between Everett worlds and alternative possibilities.

While a picture of Everett worlds as causally interacting appears natural if all of the Everett worlds are viewed as co-actualities, once Everett worlds are regarded as alternative possibilities the causal picture threatens to undermine important conceptual connections between causation, counterfactuals and modality more generally. And, to return to a theme from chapter 1, our intuitions have very little evidential force when they concern such theoretical matters as whether distinct components of the universal state can causally interact. What matters is the coherence and explanatory power of overall physics-metaphysics package deals, and a non-causal picture of quantum interference is a key part of the quantum modal realist package deal. For this reason, I suggest that despite the unfamiliarity of the non-causal picture of quantum interference it ought to be preferred to the causal picture.

As I have set it out in this section, the decoherent histories framework is non-relativistic, but this is merely a simplifying assumption. Instead of the non-relativistic projectors described above, we can work in terms of a preferred basis corresponding to field values; probabilities and the consistency conditions can then be specified in terms of path integrals.[8] This seamless extension of EQM to quantum field theories is seen by proponents as one of the strongest arguments in favour of EQM (Wallace 2012).

2.4 Overlap vs Divergence

This section discusses two different ways to interpret the decoherent histories formalism. *Overlapping EQM* allows that macroscopic objects and events may be part of several different Everett worlds; *diverging EQM* maintains that macroscopic objects and events are always *worldbound*, each being part

[8] For further details, see Gell-Mann and Hartle (1990).

of one Everett world only. I argue that this contrast comprises a case of metaphysical underdetermination, and that (following the methodology sketched in section 0.2 of the introduction) we need to make our choice between overlap and divergence on grounds of broader theoretical utility.

Now we can turn to the case of apparent underdetermination which the decoherent histories formalism raises. It concerns the metaphysics of Everett worlds, those entities that are represented by individual decoherent histories. Are Everett worlds mereologically distinct from one another?—i.e., do they overlap? According to the *overlapping* interpretation of the decoherent histories formalism, Everett worlds have initial segments in common. According to the *diverging* interpretation of the formalism, Everett worlds may have qualitatively indiscernible initial segments, but they are always mereologically distinct with no segments in common. The question can be glossed in terms of the most appropriate metaphor for Everett worlds. Which is correct: to talk of 'parallel worlds' or of 'splitting worlds'? Both images are common in popularizations of the many-worlds idea. But since they are jointly inconsistent, from a broadly scientific realistic perspective only one of the images can accurately capture the structure of an Everett multiverse. To adopt them both as literally accurate would be to incorporate a contradiction into the foundations of EQM.

Why has there tended to be a presumption of overlap in most versions of EQM? Consider the following pair of histories, both members of a decoherent history space:

$$
\begin{aligned}
C_{\underline{\alpha}} &= \widehat{P}_{\alpha_n}(t_n)\,\widehat{P}_{\alpha_{n-1}}(t_{n-1})\ldots\widehat{P}_{\alpha_0}(t_0) \\
C_{\underline{\alpha}'} &= \widehat{P}_{\alpha'_n}(t_n)\,\widehat{P}_{\alpha'_{n-1}}(t_{n-1})\ldots\widehat{P}_{\alpha'_0}(t_0)
\end{aligned}
$$

Suppose that $\alpha_k = \alpha'_k$ for all k except k=n. Then the two histories agree at all times up to t_{n-1} but differ at t_n. According to overlap, the entities represented by operators $\widehat{P}_{\alpha_0}\ldots\widehat{P}_{\alpha_{n-1}}$ in C_α are numerically identical to the entities represented by the projection operators $\widehat{P}_{\alpha'_0}\ldots\widehat{P}_{\alpha'_{n-1}}$ in $C_{\underline{\alpha}'}$. According to divergence, the entities concerned are (numerically distinct) qualitative duplicates. The quantum formalism itself does not settle the matter: the metaphysics of quantum theory is in this sense underdetermined by the physics.

What is at issue here is a metaphysical question about the way that projection operators represent reality. If projection operators represent token property-instantiations, then—because the projection operator is part of many histories—the same token is part of many worlds, and overlap follows. If projection operators represent types of property instantiation, then overlap does not follow, since the types may have many tokens. There is therefore clear scope for combining divergence of worlds with the decoherent histories formalism.

Treating projection operators as representing types of property-instantiation is not the only option for implementing divergence. We can also model the mode of representation of property-instantiations as irreducibly plural: on this view, a single projection operator represents (plurally) some property-instantiations. This interpretation of the formalism, too, is fully compatible with divergence of worlds; and it might be more attractive to nominalists, since it brings with it no ontological commitment to types.[9]

The upshot of this discussion is that the decoherent histories formalism is neutral between divergence and overlap of worlds. The choice between these two metaphysical pictures is underdetermined by the physical considerations so far adduced. Notice that neither of the two suggestions I have made for how to ground a diverging ontology of worlds alters the physics of EQM at all. Divergence, as I understand it, involves a distinctive way of aligning the physics of decoherent histories with the metaphysics of Everett worlds.

The crucial difference between diverging EQM and Deutsch's 1985 proposal (discussed in section 2.1) lies in the way in which the diverging ontology is grounded. In the diverging version of EQM outlined here and defended by Saunders (2010b) and by A. Wilson (2012), the diverging metaphysic arises from a distinctive choice of correspondence rules, or bridge principles, connecting the fundamental state with the multiverse. In Deutsch's proposal, the diverging metaphysic arises from directly modifying the fundamental ontology of the theory—that is, from adding to the fundamental state description. What I'm suggesting, and what I take Saunders to be suggesting, is to treat coarse-grained consistent histories as corresponding individually to diverging possible worlds, rather than, with Deutsch, treating coarse-grained consistent histories as corresponding to infinite sets of diverging possible worlds. In Deutsch's 1985 picture, it is always determinate that there exists a continuously infinite number of worlds. In my envisaged picture, it may be indeterminate how many worlds there are. But such vagueness seems in any case inevitable once the idea of an exact 'interpretation basis'—which features prominently in Deutsch's proposal—

[9] Thanks to L.A. Paul for discussion of this point.

is abandoned. (See chapter 5 for further discussion of indeterminacy in number and nature of Everett worlds.)

The question of whether worlds overlap or diverge—of whether the entities represented by equivalent projection operators appearing in the histories in question are numerically or merely qualitatively identical—appears coherent and substantive, but still it has a distinctively metaphysical flavour. We might therefore do well to look to metaphysics when answering it. In the next section, I apply an influential argument of David Lewis's to the Everettian context and use it to argue in favour of a diverging interpretation of EQM.

2.5 The Future Argument for Divergence

How are we to arbitrate between divergence and overlap? Here, metaphysics has something to offer to the debate. A parallel dialectic between overlap and divergence crops up in the context of Lewisian modal realism, and a prominent argument due to David Lewis supports divergence over overlap. Lewis (1986b) opted for a diverging version of Lewisian modal realism, as a corollary of his general rejection of so-called 'trans-world identity'. Famously, it is a part of Lewis's counterpart theory that Hubert Humphrey and his counterparts in other worlds are numerically distinct, no matter how similar. Likewise, Lewis held that the actual world and its counterpart worlds are numerically distinct, no matter how similar.

The general Lewisian argument against trans-world identity is the 'argument from accidental intrinsics'. If an object has an intrinsic property in one world and lacks it in another, then it both has and lacks the property. Making sense of this apparently requires turning intrinsic properties into relations to worlds.[10] But, Lewis maintains, intrinsic properties such as shape are monadic properties; they're not relations to anything.

The argument from accidental intrinsics applies to cases of overlapping Everett worlds as follows. Consider an overlapping set of Everett worlds, and ask what an agent embedded in them should think about an upcoming

[10] Non-relational approaches are possible, at the cost of introducing additional complications. For example, we might employ a primitively modalized copula 'is', or a multitude of different copulas bound to different worlds. For simplicity I will suppress these complications.

branching event. Suppose there will be a sea-battle in one world, but no sea-battle in the other. Then there is no way things will be *simpliciter*; there is only a way things will be according to each world. Given overlap, the property of *being such that there will be a sea-battle* is not a monadic property (as on the divergence view) but a relation to an Everett world.

The Lewisian objection to overlap, then, is that it makes unavailable unrelativized future-directed contents for our thought and talk. This objection applies to overlap in EQM just as much as to overlapping modal realism. In overlapping EQM, future-directed predications are turned into relations to histories, and agents can stand in many such relations. The electron (prior to measurement) *will be spin-up* relative to one Everett world and *will be spin-down* relative to another. Lewis charged that this made nonsense of our thought about the future; I wish only to defend the more modest claim that these considerations break the theoretical underdetermination between overlap and divergence. Divergence involves less significant revision of our ordinary thought about the future, and diverging EQM is consequently to be preferred over overlapping EQM. This is the Future Argument for overlap.

In other work (A. Wilson 2011, 2012) I have considered and rejected two ways of defending overlapping EQM from the Future Argument: the Lewisian metaphysics of continuant objects discussed by Saunders & Wallace (2008), and the branching-time semantics defended by Belnap, Perloff, & Xu (2001). The Lewisian metaphysics does not provide a complete solution to the problem, since it fails to provide suitable contents for propositions concerning events after the speaker's own death. Branching-time semantics fails to solve the problem at all: its proponents gesture at a pragmatic solution to the problem (which they call the 'Assertion problem', but which in fact applies far more widely than to speech acts of assertion), but I have argued elsewhere (A. Wilson 2012) that this solution is both underdeveloped and inadequate.

Lewis had no hesitation in endorsing divergence of possible worlds as a response to the problems generated by overlap. Divergence gives us appropriate contents for our future-directed thought and talk: when we ask if there will be a sea-battle, we are asking if there is a sea-battle located in the future of the one and only Everett world that we ourselves occupy. Given overlap, we are part of many Everett worlds, some of which do and some of which do not contain a sea-battle; given divergence, we are part of only a single Everett world and so there is a straightforward fact of the matter—whether that world contains a sea-battle—about which we can have beliefs, make assertions, and so forth.[11]

Considerations from one area of metaphysics (the metaphysics of future-directed mental and semantic content) are here being used to constrain another (the ontology and mereology of worlds.) Divergence does involve commitment to more token property-instantiations than overlap; however this kind of parsimony is quantitative parsimony, which Lewis famously denigrated. The extra property-instantiations that overlap brings are just the same sort of things already recognized by defenders of overlap, so Everettians who take this stance may legitimately endorse divergence of Everett worlds as a response to the problems with overlap.

While I take the need to make sense of future-directed contents in the theory of semantic and mental content to be compelling reason to adopt diverging EQM, the need to make sense of objective future-directed probabilities is even more decisive, because of the centrality of questions about objective probability to quantum mechanics. Overlap undermines ascriptions of objective probabilities to future events, since it does not provide any suitable propositional contents to which these probabilities can attach. The entire body of evidence we have for quantum mechanics is probabilistic in nature; the scientific applications of quantum theory would grind to a halt without the ability to assign probabilities to future outcomes. It would be absurd for metaphysicians to try to tell physicists that EQM is unworkable, on the grounds that it is incompatible with the existence of objective probabilities for future events. Instead, it is incumbent on metaphysicians to find a metaphysical picture that can accommodate the physics of EQM unrevised. These considerations will be further developed in chapter 3.

We can now see how the debate between overlapping EQM and diverging EQM maps onto the methodological framework set out in section 0.4 of the introduction. While there is an apparent underdetermination in the ontology of EQM when we look at the decoherent histories formalism in isolation, this underdetermination can be broken by considerations from the wider theoretical context involving assignments of contents—and objective probabilities—to propositions in general, and in particular to propositions about the future. What the methodology of section 0.4 of the introduction

[11] Tappenden (2008) objects to divergence-style EQM by challenging the possibility of situated agents referring to one divergent world as opposed to another. For a response, see Saunders & Wallace (2008).

tells us is to compare complete physics–metaphysics package deals. The combination of overlapping EQM with eliminativism about future-directed contents (and probabilities thereof) is set against the combination of diverging EQM with a self-locating conception of future-directed contents. The former package deal has problems with future-directed contents—and therefore, crucially, with future-directed objective probabilities—that the latter lacks. And if we grant that the best physics–metaphysics package deal in the vicinity incorporates a diverging metaphysics, then interpretational metasemantics licenses the conclusion that divergence is *ipso facto* correct.

One way to resist the argument of this section would be to cling on to an overlappingmetaphysics and to make corresponding adjustments to the semantics. This is the approach of Saunders & Wallace (2008)[12] and of Wallace (2012: ch.7). In criticizing this semantic approach, I appealed (A. Wilson 2012) to the physics/metaphysics evidential asymmetry in combination with a related asymmetry: the linguistics/metaphysics evidential asymmetry. Like physics, linguistics (in particular: natural language semantics) is a mature science, with an empirical track record more impressive than the track record of analytic metaphysics. Accordingly, if our choice is between modifying basic principles of semantics and modifying basic principles of metaphysics, then the metaphysics is what has to go.

We can picture the dialectical situation as involving twin anchor points: orthodox semantic theory on one side and the physics of EQM on the other. The task of the naturalistic metaphysician is to build bridges between these anchor points in a minimally revisionary way, one that generates no undesirable knock-on effects elsewhere. The metaphysics is put in by hand, but it is the pursuit of best total theory that guides our choice of what to put in.

2.6 Defusing Motivations for Overlap

Overlap of worlds has historically played a significant part in philosophical thinking about EQM. Before we leave it behind altogether, I will review some influential considerations in favour of thinking in terms of overlap and sketch the response that I think a friend of divergence should make to such arguments. They are of two main kinds; arguments from a priori

[12] Actually, two distinct proposals are woven together in that paper; one is semantic, the other metaphysical. See A. Wilson (2011) for discussion.

modal metaphysics, and arguments from fidelity to the spirit of the physics. I will treat them in turn.

I think that the main source of the presumption in favour of an overlapping interpretation of EQM amongst philosophers of physics is this a priori assumption about modal metaphysics and its interaction with theory choice in physics, introduced in 1.1:

> *Physical Actualism:* Each model of a complete physical theory represents exactly one possible world.

Physical Actualism entails that EQM, like other physical theories, has models each of which corresponds to exactly one possible world. But models of EQM are complete multiverses, and so each model includes many histories; and histories are structures that we naturally want to think of as representing *alternatives* to one another. Physical Actualism entails that these histories represent different parts of one single possible world; it precludes thinking of each history as representing a possible world in its own right.

We can get from the view that multiverse models of EQM represent single metaphysically possible worlds to the conclusion that overlap is preferable to divergence in a couple of ways, depending on what we take to be a sufficient condition for distinctness of possible worlds. Consider the following pair of arguments against a conception of Everett worlds as diverging:

The way of spatiotemporal isolation:

1. Assume for *reductio* that distinct Everett worlds diverge rather than overlap.
2. Assume Physical Actualism.
3. (from 1,2) Some possible world contains multiple diverging Everett worlds.
4. Diverging Everett worlds are spatiotemporally isolated from one another.
5. Spatiotemporal[13] isolation is a sufficient condition for distinctness of possible worlds.
6. (from 3,4,5) Some possible world contains many distinct possible worlds. (*Reductio*)

[13] Or 'spatio-temporal or analogously spatio-temporal'. See section 1.6, as well as Lewis (1986b: 75–6).

The way of causal isolation:

1*. Assume for *reductio* that distinct Everett worlds diverge rather than overlap.
2*. Assume Physical Actualism.
3*. (from 1*, 2*) Some possible world contains multiple diverging Everett worlds.
4*. Diverging Everett worlds are causally isolated from one another.
5*. Causal isolation is a sufficient condition for distinctness of possible worlds.
6*. (from 3*, 4*, 5*) Some possible world contains many distinct possible worlds. (*Reductio*)

In order to give coherent criteria of individuation for possible worlds, it would seem that we have to adopt either thesis 4 or thesis 4*[14]; it has proved difficult to come up with any other suitable sufficient condition for distinctness of possible worlds. And most philosophers who have thought about the interaction of modal metaphysics and physical theory accept Physical Actualism. So most philosophers will accept premises which lead to problems in combination with divergence, and they will therefore find overlap a natural interpretation of EQM.

Authors that I think are moved, directly or indirectly, by considerations of this sort include Saunders (1994, 1995, 1998), Wallace (2006, 2012), Tappenden (2008), Albert & Loewer (1988), Barrett (1999), Greaves (2004, 2007b), David Lewis (1986b, 2004), Peter Lewis (2007, 2009), and perhaps Everett himself (Everett 1957a). To give just one example: 'The essential difference, now, from Lewis's modal metaphysics, is that whilst these histories all co-exist, they are not separate worlds in his sense. They are not physically closed-off from one another' (Saunders 1994). Physical Actualism is also behind the description by Greaves (2004) of the Everett interpretation as 'objectively deterministic', where determinism is taken to entail that the state of a possible world at some time fixes (along with the laws) the state of that possible world at every time. And (as I shall argue in detail in chapter 3) it is behind the 'incoherence problem', the highly influential objection to

[14] Lewis is the most famous advocate of thesis 4; see Lewis 1986b, section 1.6. No modal realists that I am aware of explicitly adopt thesis 4*, but it is the natural analogue for a modal realist of the 'Causal Criterion of Reality', which Kim (1992) traces back to Samuel Alexander (1920); see Kistler (2002) for a recent discussion.

EQM which maintains that assigning probabilities to worlds simply makes no sense when every world really exists.

Everettians should abandon Physical Actualism. It may be true of one-world physical theories that each model corresponds to a single possible world, but it need not be presupposed in the context of all possible physical theories. In particular, I think that Physical Actualism should be rejected in the context of many-world physical theories such as EQM. In chapter 1 I argued that abandoning Physical Actualism clears the way for a powerful and reductive theory of metaphysical modality. Far from being a threat to our thought about modality, the rejection of Physical Actualism allows us to improve on orthodox accounts of the nature and function of modal discourse.

Let us therefore turn to what I see as the other main argument against divergence; that divergence misrepresents the physics of EQM. It is sometimes claimed that while divergence is *compatible* with EQM, it fails to do justice to the spirit of the theory. Simon Saunders has used this style of argument in a number of papers from the 1990s. Here is one example:

> The universal state, with its unitary evolution, is a single object in its own right. We may take various cross-sections through this object, and consider relations among various of its parts, but the totality of cross-sections and relations does not exist as something over and above the original. I take it that this is the guiding inspiration of the relational approach, and the core concept of the physics. Saunders (1998)

I have argued above that the quantum formalism does not in fact distinguish between taking multiple instances of a projection operator to represent numerically distinct entities and taking them to represent numerically identical entities. If this is correct, then the formalism does not distinguish between divergence and overlap. This conclusion is one that Saunders has come to accept. In his more recent writings, he argues that divergence is the more serviceable metaphysics—the metaphysics which best meshes ordinary language and thought with the physics of quantum mechanics—and hence that it is, *ipso facto*, the correct metaphysics:

> [T]he suspicion is that whether worlds in EQM diverge or overlap is underdetermined by the mathematics. One can use either picture; they are better or worse adapted to different purposes. If so, it is pretty clear which is the right one for making sense of uncertainty. Saunders (2010a)

Saunders diagnoses the presumption of overlap as a historical accident, deriving from Everett's use of the term 'branching' to set out his theory. I suspect that the 'misrepresenting the spirit of the physics' argument also reflects an intuitive, or aesthetic, preference for overlap. There is after all a real sense in which overlap is more ontologically minimal than divergence.[15] But while I do recognize the appeal of the overlapping conception, intuitive and aesthetic considerations are of very limited value when it comes to interpreting a theory as unfamiliar, and with such profound implications, as EQM.

2.7 Summary

This chapter has laid out and defended the most important features of diverging EQM, a package deal with both physical and metaphysical components that is applied in the rest of the book to problems in the metaphysics of chance and modality. These are its key features:

- *Conservatism:* Diverging EQM is a straightforwardly realist approach to the quantum formalism. It explains the effectiveness of the quantum algorithm without supplementing it with hidden variables, with dynamical collapse mechanisms, with stochastically evolving minds, or with a world-splitting mechanism.
- *Decoherence:* Diverging EQM relies on the well-understood dynamical process of decoherence to individuate worlds. It does not need the stipulation of an 'interpretation basis' to determine a set of privileged physical variables.
- *Emergence:* Everett worlds in diverging EQM are derivative: the fundamental reality is the universal state. Macroscopic worlds, objects, and events are all emergent phenomena.
- *Divergence:* According to diverging EQM, Everett worlds are mereologically distinct from one another; worlds never have parts in common. Everett worlds are represented by complete decoherent histories, but projection operators appearing in distinct histories never represent

[15] Moreover, since the additional ontology is all at the derivative level, some popular formulations of the principle of theoretical parsimony do not tell against it at all. See Schaffer (2014) for discussion.

numerically the same entity. Each Everett world inhabits its own spacetime, and no Everett world has any parts in common with any other world.

Diverging EQM, characterized by these four principles, vindicates the metaphor of 'parallel worlds' over the metaphor of 'splitting worlds' or 'branching worlds'. In chapter 3, I explain how objective chance fits into (and further motivates) diverging EQM.[16]

[16] Some material from this chapter has been reproduced from my previous articles in *The British Journal for the Philosophy of Science*: 'Objective Probability in Everettian Quantum Mechanics' 64(4), December 2013, 709–37; 'Everettian Confirmation and Sleeping Beauty' 65 (3), September 2014, 573–98; 'The Quantum Doomsday Argument' 68(2), June 2017, 597–615.

3
Emergent Chance

3.1 Introduction

Objective probability has always been a focus for criticism of EQM. Opponents of the theory who might be prepared to grant the claims it makes about the ontology of Everett worlds often argue that EQM fails to give these worlds an appropriate connection to quantum probability. In this chapter I offer a new positive theory of objective probability—*chance*—in the Everettian setting, a theory which depends on distinctive features of quantum modal realism.

In chapter 2 I argued that Everettians face a key theoretical choice when formulating the ontology of EQM: they must decide between overlapping Everett worlds, which share initial segments, and diverging Everett worlds, which do not. Is the multiverse made up of splitting worlds or of parallel worlds? The Future Argument of section 2.5 shows that overlapping EQM prevents us from coherently assigning chances to alternative future outcomes; so diverging EQM earns its keep in the service of chances. Even so, adopting diverging EQM does not suffice to make full sense of probability in the Everettian picture. Further metaphysical ingredients—bridge principles connecting the physics of EQM with the metaphysics of modality—are required for a satisfactory overall theory of Everettian chance that is hospitable to non-trivial chances attaching to the outcomes of future measurements.

Section 3.2 surveys objections to EQM based around objective probability, dividing them into i) challenges to the coherence of Everettian probability, ii) challenges to Everettian probabilities being specified by the Born rule, and iii) challenges to the possibility of EQM being probabilistically confirmed by evidence. The challenge to coherence is the most pressing: in section 3.3 I resolve it through a quantum modal realist realignment between the physics of EQM and the metaphysics of modality. The bridge principles invoked in this realignment combine with the physics of diverging EQM to comprise Indexicalism, a complete framework for understanding Everettian probability. In sections 3.4, 3.5, and 3.6 I argue that Indexicalism is indispensable for making sense of chance in the Everettian picture. I offer

The Nature of Contingency: Quantum Physics as Modal Realism. Alastair Wilson, Oxford University Press (2020).

DOI: 10.1093/oso/9780198846215.001.0001

three arguments for Indexicalism: the first (the Qualitative argument) turns on general conceptual truths about the nature of chance, the second (the Quantitative argument) turns on the role of the Born rule in the quantum algorithm, and the third (the Epistemic argument) turns on the need for EQM to be confirmable and disconfirmable by evidence in a coherent and non-pathological way. Sections 3.7 and 3.8 flesh out the Indexicalist conception of Everettian chance, showing how it plays the various metaphysical roles that recent metaphysicians have taken to be constitutive of chance, and exploring the novel properties that result from its status as emergent. Taken together, this chapter offers a complete metaphysics for chance in the Everettian context, licensing the move from diverging EQM to the full quantum modal realist framework.

3.2 The Everettian Probability Problem(s)

The move from overlap to divergence is an important first step in the rehabilitation of Everettian probability, as I argued in section 2.5. Unlike overlapping EQM, diverging EQM includes straightforward contents to which objective chances can attach: probabilities are probabilities of self-location within some set of Everett worlds. Trouble, however, remains just around the corner. This section introduces the most serious objections to EQM that arise from considerations of objective probability, and shows how these objections recur in the context of diverging EQM. I briefly survey the *fission programme*, which aims to make sense of quantum mechanics without the concept of chance, and describe how it can be adapted to divergence. I then argue that Everettians should pursue a different approach to probability, one that recaptures chance rather than eliminating it altogether.

The 'probability problem' in EQM has been much discussed, but the threads of the debate can be spun out into three distinct components:[1]

- *The Incoherence problem:* In EQM, the fundamental laws of nature are deterministic. An initial quantum state, in combination with unitary evolution, determines a complete multiverse. So, what can objective probability possibly amount to in a theory like this?

[1] The terms 'Incoherence problem' and 'Quantitative problem' are due to Wallace (2003b); the term 'Epistemic problem' is due to Greaves (2007a). I divide things up slightly differently from these authors.

- *The Quantitative problem:* In EQM, quantum amplitudes are physical quantities. There seems to be no a priori reason why these quantities in particular should guide rational credence. So, what explains the central role of the Born rule in the quantum algorithm?
- *The Epistemic problem:* In EQM, every physically possible sequence of measurement outcomes is observed in some Everett world, including sequences with very low amplitude. So, how can statistical empirical evidence ever confirm or disconfirm quantum theories?

The Incoherence problem has historically been the focus of discussion of Everettian probability, and it is easy to see why. All physically possible sequences of events in fact occur in the Everett multiverse, and so it seems inevitable that they should all be assigned objective probability 1. Physically impossible sequences of events do not occur; those should each be assigned objective probability 0. So what could non-trivial chances possibly amount to in the Everettian setting?

One prominent answer to this question says: *nothing at all.* Proponents of the *fission programme*, such as Hilary Greaves and Paul Tappenden, deny that objective probability makes any sense within EQM, and seek to explain away the appearances as of objective probabilities. According to the fission programme, it makes no sense to attribute non-trivial objective probabilities to future outcomes—after all, each such outcome will definitely occur in some Everett world—but it does make sense to *care differentially* about these future outcomes. Branches that have a high Born-rule weight are not in any sense more probable; however, according to proponents of the fission programme, such branches are nevertheless more deserving of our practical concern. If the world splits, we may expect to have multiple *descendants* in multiple worlds, each of which is physically and psychologically continuous with our present selves; and we should care more about our descendants in Everett worlds with high weight than we should care about our descendants in Everett worlds with low weight.

Despite its name, the fission programme is compatible with both overlapping and diverging conceptions of EQM. Even if I know I am located within a single Everett world in a diverging multiverse, it might still make sense for me to care differentially about how things turn out in the various other Everett worlds according to their weights, and accordingly to have preferences over which measured sets of Everett worlds to bring into being through my actions. However, the rationale for the fission programme is much weaker in the context of diverging EQM than it is in the context of overlapping EQM. When motivating the fission programme, Greaves relies on the fact that the agents in various future Everett worlds amongst which we are to distribute our concern are our descendants in those worlds, pointing out that such agents bear the usual sort of relations (causal, informational, etc.) to us that are usually taken to support judgements of personal identity or gen-identity. Once these links are severed, as they are in the diverging picture, it is no longer obvious why it should be rational to care about these other agents at all, let alone to care about them differentially according to the branch weights. Divergence accordingly saps the motivation for the fission programme.

Regardless of the question of divergence vs overlap, I suspect that the fission programme does too much damage to our ordinary conception of the world to be a part of the best Everettian package deal of physics plus metaphysics. We can draw an analogy between the fission programme and Arthur Eddington's famous proclamation (Eddington 1928) that physics has discovered that tables are not solid, being composed of atoms with large stretches of empty space between them. Although the atomic theory of matter is now firmly embedded in our twenty-first century worldview, nobody has stopped talking about tables—for perfectly good reasons. Nobody is likely to stop talking about chances anytime soon either—for equally good reasons. A better reaction to the modern physics of matter is to reassess our view of the nature of solidity, so that tables continue to count as solid despite atomic theory's revelations about their constitution. Likewise, a better reaction to the physics of EQM is to reassess our view of the nature of objective probability, so that the world continues to count as objectively probabilistic despite our discovery of the Everettian multiverse.

We can, and should, move beyond the fission programme. Everettians do better to naturalise probability within the Everettian picture, rather than to eliminate probability altogether. Of course, it remains to be seen if any adequate non-eliminative account is available. In section 3.3, I provide my own positive account of Everettian probability; I then go on to defend the account in sections 3.4, 3.5, and 3.6. If my picture can be sustained, then it is preferable to the fission programme, since it does much less damage to our pre-Everettian worldview.

3.3 The Ingredients of Indexicalism

The ontology of a many-worlds physical theory may be aligned with the ideology of modal metaphysics in multiple ways. In this section I present a set of 'bridge principles' connecting the physics of EQM with the metaphysics of modality.

Recall from section 1.1 that the following principles make up quantum modal realism:

- *Diverging EQM:* Everett worlds do not overlap; each macroscopic object and event exists in one Everett world only.
- *Individualism:* Distinct Everett worlds comprise alternative metaphysical possibilities.
- *Propositions-as-sets-of-worlds:* Contingent qualitative propositions are sets of Everett worlds—a proposition P is true at an Everett world w if and only if w is a member of P.
- *Indexicality-of-Actuality:* Each Everett world is actual according to its own inhabitants, and only according to its own inhabitants.
- *Everettian Chance:* The objective chance of an outcome is the quantum-mechanical weight (squared amplitude) of the set of Everett worlds in which that outcome occurs.

In this chapter, the last of these principles—Everettian Chance—is what is at stake. The first four principles, when combined with diverging EQM, comprise a package deal that I call Indexicalism. In the remaining sections of this chapter, Indexicalism will be exploited to solve the incoherence (3.4), quantitative (3.5), and epistemic (3.6) problems with Everettian probability. Sections 3.7 and 3.8 parlay these solutions into a justification for Everettian Chance, thereby completing the quantum modal realist framework.

I shall begin by considering and deflecting some initial objections to the bridge principles with which Indexicalism supplements Diverging EQM. Take first the principle Individualism, and recall from section 1.1 that Individualism's chief rival is Collectivism:

Individualism: If X is an Everett world, then X is a metaphysically possible world.

Collectivism: If X is an Everett multiverse, then X is a metaphysically possible world.

I take it that neither Individualism nor Collectivism enjoys direct intuitive support. They both contain distinctively metaphysical technical terms, and it is implausible to think that intuition can arbitrate between principles including such terms. However, Collectivism has been the default position amongst philosophers who have thought about EQM; why might this be?

The presumption of Collectivism, I suspect, derives from a more general assumption about the relationship between physics and metaphysics, already discussed in chapters 1 and 2: Physical Actualism. Here is a reminder of what this principle says:

> *Physical Actualism:* Each model of a complete physical theory represents exactly one metaphysically possible world.

I do not want to suggest that everyone who—implicitly or explicitly—adopts Collectivism must be committed to Physical Actualism. Other reasons for adopting Collectivism might involve the presumed causal or spatiotemporal connectedness of Everett worlds which feature as independent premises of the arguments against divergence considered in section 2.5. But I do think that Physical Actualism is lurking in the background of many discussions.[2]

Physical Actualism, while a natural assumption in the context of one-world physical theories such as classical mechanics, should be rejected in the context of many-world physical theories such as EQM. Physical Actualism entails that each model of EQM represents exactly one possible world. But models of EQM are complete multiverses; each model includes multiple histories, structures which are naturally treated as representing *alternatives* to one another. Physical Actualism entails that these histories represent different parts of one single possible world, so it precludes thinking of each history as representing a possible world in its own right.

It may be illustrative to consider a hypothetical physical theory that has as a model the Lewisian plurality of worlds (Lewis 1986b). Physical Actualism entails that this plurality must correspond to a single possible world; thus taking Physical Actualism to be a priori true requires taking the Lewisian modal realist account of modality to be a priori false. Since Lewisian modal realism is not a priori false, Physical Actualism is not a priori true. But there are no a posteriori grounds for Physical Actualism either; so we are left with no good reason to adopt it.

Nothing in the physics of quantum mechanics, or in a priori metaphysics of modality, compels us to adopt either Individualism or Collectivism. That

[2] Tim Maudlin gives a particularly clear statement of Physical Actualism in a different context: 'Let us suppose (and how can one deny it) that every model of a set of laws is a possible way for *a world governed by those laws to be*' (Maudlin 2007: 67; emphasis in original).

choice should be made instead on grounds of overall theoretical utility, according to the methodology sketched in section 0.3 of the introduction. I will argue that an application of that methodology provides us with strong reasons to prefer Individualism to Collectivism. In the remainder of this section, I supplement diverging EQM with Individualism and with some additional principles relating metaphysics to the physics of EQM; this package of views is then applied to the probability problem in subsequent sections.

The first additional ingredient in the Indexicalist package is an identification of ordinary, contingent, qualitative propositions—the sort of propositions that have non-trivial objective chances—as sets of Everett worlds. An initial plausibility argument for this identification goes as follows:[3]

1. Objective chances attach to propositions.
2. Branch weights attach to sets of Everett worlds.
3. Objective chances are branch weights.
4. Propositions bearing objective chances are sets of Everett worlds.

My treatment of chance-bearing propositions as sets of Everett worlds obviously has much in common with Lewis's treatment of propositions as sets of Lewis worlds. As with the Lewisian account, the account of chance-bearing propositions as sets of Everett worlds is in principle compatible both with conceptions of actuality according to which it is an absolute property (see e.g. Bricker 2001) and with the Lewisian conception of actuality as indexical in nature (see e.g. Lewis 1970, 1986b). The latter conception fits much more naturally with EQM. Although Everett memorably insisted (in a note added in proof to his original article) that all branches were actual, it is clear from the context that he took the terms 'actual' and 'real' to be synonymous with one another: 'From the viewpoint of the theory *all* elements of a superposition (all "branches") are "actual", none any more "real" than the rest"' (Everett 1957a); the same sentence appears in Everett (1957b). It is entirely unsurprising that Everett did not consider the possibility of an indexical conception of actuality; this idea was only introduced (into academic philosophy) by Lewis, over a decade later. Now that we have such a conception, there is no reason to abjure it on Everett's authority.

In what follows, I will argue that indexical actuality is a key component in a satisfying Everettian response to the probability problem. For a

[3] This formulation was suggested to me by Cian Dorr.

proposition P to be true at the actual world, in the Indexicalist picture that I will develop, is for the actual Everett world to be a member of the set of Everett worlds P. This account works regardless of whether P is picked out in qualitative or non-qualitative terms.

This completes my initial presentation of the ingredients of Indexicalism. Now, why should we accept the Indexicalist package?

3.4 The Qualitative Argument for Indexicalism

The first argument I offer for Indexicalism is that it solves the Incoherence problem with Everettian probability, and hence that Indexicalism allows us to make sense of non-trivial objective probabilities in a many-worlds scenario.

At the heart of the Incoherence problem is the question of whether, in EQM, we can coherently assign non-trivial[4] objective probabilities to propositions, as we do in the context of an orthodox one-world metaphysics of objective probability. Probabilities are necessarily probabilities of propositions. But, so the objection from the Incoherence problem runs, given any quantum state, and the deterministic unitary state evolution, the future will be described, with objective probability one, by a highly specific proposition which entails the existence of countless Everett worlds. The total state of reality at any time and the laws of nature therefore jointly determine the whole of the multiverse, with probability one. All propositions entailed by this maximal proposition have probability one; all propositions inconsistent with it have probability zero. Objective probability is trivialized.

This objection has been around for a long time, and it has been made eloquently by Barry Loewer, amongst others:

> How are we to understand this measure? What does the measure measure? None of the familiar notions of probability seem appropriate. Clearly it cannot be construed as measuring the chances in a stochastic law... since [EQM][5] is a deterministic theory. Loewer (1996: 230)

[4] By 'a non-trivial objective probability' I mean an objective probability other than zero or one.

[5] Loewer actually addresses this criticism at the 'instantaneous minds view', a version of the many-minds theory due to Lockwood (1989). But the criticism applies with equal force to decoherence-based EQM.

According to the Incoherence Problem, weights—which attach to individual Everett worlds (and sets thereof) and only degenerately to whole multiverses—are simply the wrong type of quantity to play the role of objective probabilities.

David Wallace describes (without endorsing) a swift and dismissive response to the Incoherence problem:

> [F]ormally speaking the measure defined by mod-squared amplitude on any given space of consistent histories satisfies the axioms for a probability. Indeed, mathematically the setup is identical to any stochastic physical theory, which ultimately is specified by a measure on a space of kinematically possible histories. Wallace (2010b: 228)

Unfortunately, this response is too swift, and critics of Everettian probabilities will not be moved by it. According to the line of thought which drives the Incoherence problem, weights are categorically *not* equivalent to a measure on a space of kinematically possible histories. From the critics' perspective, any initial condition combined with the unitary evolution gives only a single (rather oddly shaped) kinematically possible history—the multiverse itself.

Collectivism is at the root of the Incoherence Problem. Since, according to Collectivism, distinct Everett branches are not alternative possibilities but co-actualities, objective probabilities must be probabilities of ways that the entire multiverse might go, rather than probabilities of ways that individual Everett worlds might go. Collectivism therefore undercuts any quick attempt to bypass the Incoherence problem. If we are to identify Everett worlds with kinematically possible histories, we must instead adopt Individualism instead of Collectivism.

Individualism is not the only component of Indexicalism that is required to provide a proper response to the Incoherence problem. Since the bearers of objective probabilities are propositions, we require an account of propositions according to which the squared-amplitude measure is defined over them. Such an account is provided by the principle Propositions-as-sets-of-worlds, which is incorporated into Indexicalism. Propositions bear objective probabilities in virtue of the squared-amplitude measure being defined over the sets of Everett worlds with which the propositions are identified. The probability that P is the probability that the actual Everett world is a member of the set of Everett worlds P.

The Indexicality-of-Actuality component of Indexicalism is likewise required if objective probabilities are to be assigned to physically contingent propositions in full generality. Amongst the propositions to which we assign objective probabilities are such propositions as 'the actual outcome will be spin-up'. If 'actual' applies to the entire Everettian multiverse, then the phrase 'the actual outcome' fails to refer uniquely, since the multiverse contains many outcomes. But given Indexicality-of-Actuality, 'the actual outcome' refers uniquely to the outcome in the speaker's own Everett world; consequently, tokens of 'the actual outcome is spin-up' are true in just those Everett worlds which contain spin-up outcomes, as desired.

This indexical treatment of propositions and of actuality faces a difficulty involving future contingents, which has featured prominently in the recent literature on Everettian probability[6] and which was discussed in chapter 2. On the overlapping metaphysical picture usually combined with EQM, distinct worlds have initial segments in common: agents facing impending quantum interactions are located in every world which emerges from the interaction. An agent about to conduct a spin measurement, it seems, is then located both in an Everett world where the result is spin-up *and* in an Everett world where the result is spin-down. Will the result be up or down? Everettians who accept overlap have trouble in making sense of this question: both results will occur. (Or—perhaps better?—*each* result will occur.)

The problem with future contingents has no analogue in Lewisian modal realism, according to which agents are always worldbound and worlds do not mereologically overlap. And the problem seems to undermine the use of indexical actuality to ground objective probabilities for future contingent propositions. If I am both part of a spin-up Everett world and part of a spin-down Everett world, why are the probabilities of the actual outcome being spin-up and of the actual outcome being spin-down not both equal to one? As I suggested in chapter 1, my favoured solution to the problem with future contingents is that proposed in Saunders (2010b), and in A. Wilson (2011, 2012) which draws on a picture of Everett worlds as diverging rather than as overlapping. It is for this reason that divergence has been incorporated into Indexicalism.

What the Incoherence problem shows us, I think, is that we need to adopt the Indexicalist package in order to make proper sense of the application of objective probabilities to ordinary contingent propositions in EQM. Once we do so, we can vindicate Wallace's claim that weights play the role of a measure over a space of possible histories. If we instead presuppose

[6] See, e.g., Saunders (1998), Greaves (2004), Saunders & Wallace (2008), A. Wilson (2011).

Collectivism, as critics of EQM usually do at least tacitly, then the only objective probabilities which could make sense in the context of EQM would be probabilities assigned to whole multiverses. Call this the Qualitative Argument for Indexicalism; it subsumes and extends the Future Argument for divergence.

A special case of the Incoherence problem should be mentioned, since it has received plenty of attention in the literature. It concerns in particular the connection between probability and uncertainty. While uncertainty is only one of a number of propositional attitudes one might have towards the propositions which bear objective probabilities, uncertainty is especially salient because of an apparent platitude connecting objective probability and uncertainty. It seems to be platitudinous that assigning a non-trivial objective probability to some proposition requires being uncertain about whether that proposition is true. EQM threatens to violate this platitude by (at least in principle) permitting situations where an agent knows the entire quantum state and accordingly has no residual uncertainty about how the multiverse will evolve. The platitude entails that such an agent cannot rationally ascribe objective probabilities to any future outcomes. The problem is sometimes framed by claiming that objective probability requires a notion of 'objective uncertainty', and that no such notion of objective uncertainty is available in EQM (Belnap & Muller 2010).

Responses to the worry about objective uncertainty have varied. The perception that objective uncertainty is unavailable in EQM has driven Everettians like Greaves (2004) and Tappenden (2008) towards versions of the fission programme and it led Wallace (2006, 2012), Saunders & Wallace (2008), and Belnap & Muller (2010) to develop semantic theories for future-directed discourse according to which objective uncertainty is available. Meanwhile, Papineau (2010) and Wallace (2012) have argued that Everettians should reject the troublesome platitude connecting uncertainty and objective probability; but all these options are unpalatable. Indexicalism offers an attractive alternative solution. It allows us to preserve objective probabilities in EQM without violating the platitude, by providing an adequate subject matter for objective uncertainty: self-location within the diverging Everett multiverse.

I argued in section 3.3 that there is no good a priori reason to prefer Collectivism to Individualism. Since the former but not the latter leads to difficulties in making coherent sense of objective probabilities, Everettians who wish to recapture non-trivial objective probabilities for outcomes of quantum interactions ought to prefer Individualism to Collectivism. With diverging EQM in the background, it is then a short step from Individualism to the full Indexicalist package.

3.5 The Quantitative Argument for Indexicalism

The second argument I shall offer for Indexicalism is that it helps to solve the Quantitative problem with Everettian probability. With Indexicalism on board, we are in a stronger position to link weights of Everett worlds to the cognitive role associated with objective probabilities.

Everettians have made a number of attempts to argue that probabilities in EQM (or some surrogate for probabilities, as in the caring measure of the fission programme) should be given by the Born rule. These arguments include the relative-frequency argument given by Everett himself and discussed in section 2.2, the decision-theoretic approach of Deutsch (1999) and Wallace (2003b, 2010b, 2012), and the indifference-based approach of Sebens & Carroll (2018). While (for reasons to be explained) I think none of these arguments show that setting credences according to the Born rule is rationally required for Everettian agents, the arguments do collectively make a compelling case that branch weights are the unique best candidates for the role of objective chances in EQM. If any element of reality is a chance measure in the Everettian picture, it is the distribution of branch weights.

My specific Quantitative Argument for Indexicalism turns on the justification of a particular premise in the decision-theoretic argument for the Born rule, as recently articulated by David Wallace (2010b, 2012). I argue that Wallace's premise *Branching Indifference* has a straightforward justification if we adopt Individualism; but without Individualism, Branching Indifference is implausible. A central component of Indexicalism is therefore a key part of one of the most promising solutions to the Quantitative problem with Everettian probability.

Wallace's strategy in addressing the Quantitative problem is to prove that a rational agent who knows that the quantum state has the structure that EQM attributes to it must allow weights to play the role of objective probabilities in his deliberations. This is a form of *probability coordination principle*. Such principles connect subjective probability (or *credence*) for rational agents, conceived of as a state of mind, with objective probability (or *chance*), conceived of as a state of the world. David Lewis's Principal Principle (Lewis 1980/1983) and New Principle (Lewis 1994) are the best-known examples of probability coordination principles; but there are many others, usually adapted to different metaphysical conceptions of objective probability. For example, Carl Hoefer (Hoefer 2007) argues for one such principle in the context of his 'Third Way' Humean chances.

Wallace's argument aims to establish the following conclusion: that an agent who believes EQM and who is given a choice between two quantum interactions, defined as completely specified unitary transformations of the quantum state, should choose the transformation which maximizes total utility (as far as the agent now is concerned) over all the outcomes which result after the interaction, weighted according to the squared-amplitude (weight) of each outcome. For those readers wondering about how free choices work at all in a fundamentally deterministic picture, note that Wallace is assuming a broadly compatibilist account of choice; I will follow suit. All that is required for the purposes of the argument is that the agent be able to compare the different prospective branching structures and identify a definite preference over them.[7]

Once we adopt Indexicalism, the first major obstacle to using Wallace's argument to ground objective probability in EQM—the Incoherence problem—is solved, as I argued in section 3.4. Objective probabilities attach to propositions, identified as sets of Everett worlds. Then objective probabilities are just probabilities of being located in a particular set of Everett worlds. What remains is to justify each of the premises of Wallace's argument. Wallace (2010b, 21012) argues for these premises, and I think most of his defences of them are persuasive enough independently of Indexicalism; however, there is one notable exception. A crucial assumption in Wallace's argument—and one which is rejected by some of the most influential proposed responses to the argument—is *branching indifference*. In this section, I will explain this premise and raise some objections to Wallace's defence of it; I will then provide an alternative defence of branching indifference that makes essential appeal to Indexicalism.

Here is Wallace's statement of branching indifference,[8] which is supposed to be a principle of rationality in the Everettian context:

[7] The broader question of free action in an Everett universe I set aside here. I presume that the causal isolation of Everett worlds and the effective indeterminism of laws governing individual Everett worlds (see chapter 4) makes compatibilist accounts of freedom no more or less plausible in quantum modal realism than in an indeterministic one-world physical theory.

[8] Wallace intends the terminology of 'branching' to be neutral between overlap and divergence. For the purposes of this section, I am following his usage.

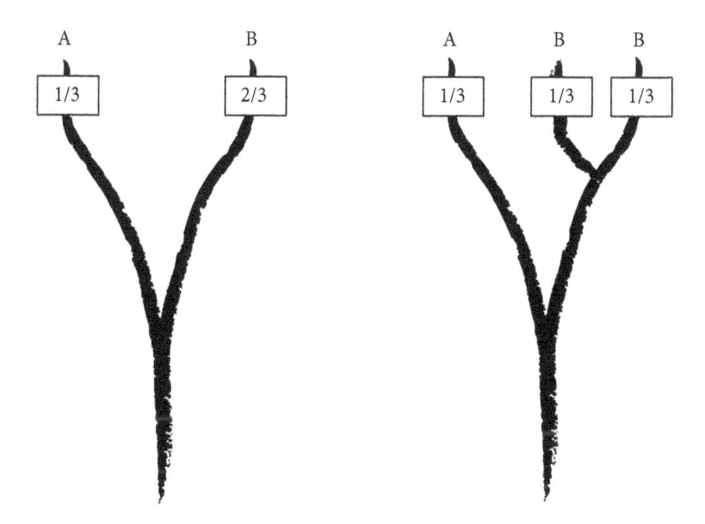

Figure 3.1. Branching Indifference in the overlap picture. After Wilson (2013a)

> BRANCHING INDIFFERENCE: An agent doesn't care about branching *per se*: if a certain measurement leaves his future self[9] in N different microstates but doesn't change any of their rewards, he is indifferent as to whether or not the measurement is performed. Wallace (2010b: 238)

To get a sense of what branching indifference involves, see Figure 3.1. Branching indifference requires a rational agent to be indifferent between the two branching set-ups illustrated. A and B are the rewards which attach to each branch; the numbers in boxes are the weights of each branch. The only difference between these set-ups is the number of branches corresponding to reward B. The range of available rewards, and their relative weights, is the same in each set-up.

Wallace's defence of branching indifference has two strands. The first strand—which I shall call the 'pragmatic defence'—denies that the number of branches is something about which we could rationally have preferences, while the second strand—which I shall call the 'non-existence defence'—maintains that the number of branches is not even well defined. After

[9] This mention of 'future selves' is dispensable. Wallace's argument applies equally to acts where some of the post-interaction branches do not contain any 'future selves' of the agent; in his decision-theoretic framework such branches can still be assigned 'rewards' by the agent at the present time.

discussing these approaches and raising some difficulties for them, I shall describe how Indexicalism grounds a third defence of branching indifference which avoids the various difficulties.

The pragmatic defence of branching indifference appeals to 'the limitations of any possible physically realizable agent': 'a preference order which is not indifferent to branching *per se* would be in principle impossible to act on: branching is uncontrollable and ever-present in an Everettian universe' (Wallace 2010b: 238). Wallace is here attempting a *reductio* of the idea that there can be a preference order which is sensitive to amount of branching. It is not clear that the *reductio* succeeds. That branching is ever-present does not prevent its being present to different degrees in different physical processes. Decoherence still holds many unknowns, and although no unproblematic scheme for quantifying degree of or amount of branching in a natural way has yet emerged, it does not seem out of the question that one might emerge in the future. Nor does it seem inconceivable that, according to some such scheme, the sort of choices we actually make might affect the amount of branching in our future.

A thorough investigation of the details of decoherence that are relevant to quantifying branching is beyond the scope of this book; but note that one of the main sources of decoherence is the operation of classically chaotic processes. In the light of this point, we might expect choices which affect the distribution of such processes across the universe to produce differential amounts of branching. In general, if choices happen to impact significantly on macroscopic variety in our future light cones, it seems at least possible that they will impact on the amount of branching in our futures.

Call the claim that there is no well-defined sense in which different physically possible acts produce different amounts of branching *branching homogeneity*. Even if branching homogeneity holds true, and our choices cannot in fact affect the total amount of branching, does it follow that we must be indifferent to branch number? Many things we value are out of our control, to a greater or lesser degree.[10] Preference orderings which are in principle impossible to act upon still appear to be coherent and possible preference orderings, pending positive arguments to the contrary. Wallace has in various writings offered positive arguments of this sort for the incoherence of unimplementable preference orders (Wallace 2003b, 2007, 2010b, 2012). Here is a representative example:

[10] Albert (2010) emphasizes this point.

> If we are prepared to be even slightly instrumentalist in our criteria for belief ascription, it may not even make sense to suppose that an agent genuinely wants to do something that is ridiculously beyond even their idealised capabilities. For instance, suppose I say that I desire (ceteris paribus) to date someone with a prime number of atoms in their body. It is not even remotely possible for me to take any action which even slightly moves me towards that goal. In practice my actual dating strategy will have to fall back on 'secondary' principles which have no connection at all to my 'primary' goal—and since those secondary principles are actually what underwrites my entire dating behaviour, arguably it makes more sense to say that they are my actual desires, and that my 'primary' desire is at best an impossible dream, at worst an empty utterance. Wallace (2007: 328)

This passage advocates a distinctive physicalist functional conception of preference, according to which an agent's preferences constitutively depend on their dispositions to action across the range of physically possible circumstances. This functional conception has the consequence that in cases where it is physically impossible for the agent's dispositions to action to be sensitive to some parameter—where that parameter is completely redundant in accounting for the agent's dispositions—then the parameter does not feature in the agent's preference ordering.

To see the consequences of the physicalist functional conception of preference for Wallace's defence of branching indifference, we must be sensitive to the distinction between the notion of a preference order over acts available to us—what we would prefer to do—and the more general notion of a preference order over future events—what we would prefer to happen. The functional conception of preference licenses us to restrict our attention to a preference order over physically possible acts. If some future event is a consequence of no possible act which is available to us, we can still have preferences defined over the event in some (perhaps wistful) sense. But it is important to see that this would not be in conflict with branching indifference, which restricts only preferences over available acts.

An agent who knows which acts are physically possible for them to perform, and who knows that each of those acts results in exactly the same amount of branching, cannot rationally prefer one act to another *on the grounds that it results in more (or in less) branching*. As a result, the assumption of branching homogeneity—that no physically possible act leads to any more or less branching than any other act—combined with the physicalist functional conception of preference, does seem to establish that compliance with branching indifference is a requirement on any physically possible rational agent.

However, if we decline to grant the assumption of branching homogeneity, then the above defence of branching indifference will not work for us. Furthermore, not only will there then be difficulties with motivating branching indifference as a rational requirement on preference orders, but there will be some prima facie reasons to think that preference orders *violating* branching indifference may in certain circumstances be rationally required. Here is the motivating thought: if a state of affairs involves significant additional branching and thereby involves the coming into existence of significantly many additional people very like myself, why is this not exactly the sort of thing I *should* care about?

Consider an analogy: suppose we have an offer from God to create multiple copies of our solar system in far-flung corners of the universe. Even if these copies would be causally isolated from us after their creation, many prominent systems of population ethics still entail that we have reason to create them (as long as their lives would be reasonably happy, of course). Why is the situation significantly different in the Everettian case, when the copies to be created are not spatially distant, but instead are in different Everett worlds? What gives the Everettian reason to value the number of duplicates of herself in her own Everett world, but not to value the number of duplicates of herself across other Everett worlds? Call this line of thought the population-ethics objection to Everettian probability.[11]

The population-ethics objection has lurked in the background of many discussions of probability in EQM. It can be traced back to Graham (1973), and more recently it has been expressed in terms of what an Everettian agent can rationally care about (see e.g. Greaves 2004, Price 2010, Kent 2010, Albert 2010). The objectors typically first point out that EQM introduces a multiplicity of the sort of people, acts, events, or artefacts to which we ordinarily ascribe normative value; they then argue that the way in which

[11] The population-ethics objection can be made most vividly by adapting Derek Parfit's 'repugnant conclusion' argument (Parfit 1984). The repugnant conclusion is that for any finite population no matter how large or how well-off its members, there will be a larger population whose members have lives only barely worth living but which is more valuable overall. Informally, quantity always trumps quality. The analogue of the repugnant conclusion in the Everettian setting is that it will generally be rational for an Everettian to induce as much branching as possible, even at significant cost, as long as enough additional people are produced with lives even barely worth living.

this multiplicity depends on our actions may have consequences for a rational agent's preference over acts.

The population-ethics objection is perhaps put most clearly by Huw Price:

> 'Where goes ontology, there goes possible preference.' Decision theory places no constraint on what agents care about, other than that it be *real*. The new ontology of the Everett view—the global wavefunction itself—thus brings in its wake the possibility of an agent who cares about *that*. Hence the challenge, in its most general form: by what right do we assume that the preferences of Everettian agents are driven by 'in branch' preferences *at all*? Price (2010: 380); emphasis in original

Of course, the Born rule strategy that Wallace is attempting to prove allows that agents can sometimes care about global properties of the quantum state: the strategy is itself a rule for choosing between sets of branches generated by different unitary transformations. So perhaps it is not quite right to say that Wallace's argument assumes that Everettian agents are driven only by in-branch preferences. But branching indifference does at least rule out caring about one particular global property of the quantum state—branch number. Price's complaint seems to apply in full force here: if there is such a thing as branch number in the Everettian ontology, how can there be something incoherent or irrational about an agent who cares about it?[12]

The combination of a physicalist functional conception of preference and the assumption of branching homogeneity provides a tenable way of ruling out preferences about branch number. To adopt the pragmatic defence of branching indifference is to insist that no physically possible agent's behaviour could ever make it correct to ascribe a preference order to the agent that is sensitive to branch number. Still, both of the components of this defence of branching indifference are very controversial; and it would be better not to have to rest much weight on it. So I turn next to consider the other strand of Wallace's defence of branching indifference.

[12] Variants of the population-ethics objection interfere with the decision-theoretic strategy in other ways. Price (2010) suggests that even if Everettians must be indifferent to branch number, they need not be indifferent to outcome number—that is, to the number of different types of Everett world diverging from an interaction. This version of the objection instead challenges Wallace's 'diachronic consistency' axiom.

The main defence of branching indifference that can be found in Wallace's writings is that branch number is not well defined, and accordingly that it cannot be a locus of normative value:

> There is no such thing as 'branch count':... the branching structure emergent from unitary quantum mechanics does not provide us with a well-defined notion of how many branches there are. All quantum mechanics really allows us to say is that there are *some* versions of me for each outcome. Wallace (2010b: 255); emphasis in original

This response can be combined with a wholesale rejection of the meaningfulness of talk about branch number;[13] alternatively, it can be combined with the thought that ascriptions of branch number, while meaningful in some minimal sense, suffer from presupposition failure.[14] If the former route is taken, ascriptions of branch number are meaningless; if the latter route is taken, they are either meaningless or false. Either way, they are never true. So, this strand maintains, once we properly grasp how branch number is an artefact of the mathematics, we will realize that there is simply no coherent way of incorporating it into our utility function.

In chapter 5 I propose an alternative treatment for ascriptions of branch number, which takes it to be *bivalently indeterminate*. Claims about branch number are always either true or false, but they need not always be determinately true or determinately false. This approach is compatible with various interpretations of the 'determinately' operator. Prima facie, the most familiar theories of vagueness—semantic indeterminacy, epistemicism, and ontic vagueness—are all live options for Everettians. If such a treatment of branch number is rejected, and the unpalatable view that claims about branch number are never true is embraced, then I think that branching indifference does follow. But the cost of this move is high. It threatens to destroy the fragile grip we have on the worldview provided by EQM, by undermining the images of overlap and divergence that seem indispensable in explicating it. Indeed, the claim that questions such as 'how many people

[13] As by Saunders (2005), whose operational derivation of the Born rule was explicitly premised on the stability of probabilities under changes in coarse-graining, and hence under changes in branch number.

[14] The classic examples of presupposition failure are cases of failed definite description, such as 'the current king of France is bald.' Such sentences need not be taken as meaningless; on a Russellian theory of descriptions they are false, for example.

are there, unrestrictedly speaking?' has no true answer seems dangerously close to a *reductio* of any version of EQM which entails it. There is accordingly good reason to look at alternative routes to justifying branching indifference.

An indeterminacy-based treatment of branch number does not by itself provide support for branching indifference. It might be tempting to think that when we know that it is indeterminate what the value of some quantity is, it is rationally required for us to be indifferent to that value. But this is not so according to precisificational conceptions of indeterminacy, which allow us, for example, to determinately prefer being non-bald to being bald even if baldness admits of borderline cases. The point can be put as follows in the case of branch number: even if branch number is indeterminate, it is determinately greater than one. Wallace tacitly acknowledges this, writing of 'some versions of me' (note the plural!) for each outcome. This limited determinacy is already enough to generate a problem for branching indifference. If, *ceteris paribus*, copies of ourselves are to be valued, then a situation which determinately contains many copies of ourselves is, *ceteris paribus*, (determinately) more valuable than one which contains fewer, even if it is indeterminate *how many more* copies the former contains than does the latter. This result follows straightforwardly from the logic of the determinacy operator, whether we appeal to epistemicism, to supervaluationism, or to any other bivalence-preserving precisification-based theory of vagueness to cash it out.

We might think that, regardless of the general metaphysical concerns we might have aboutthe 'no-such-thing-as-branch-number' response, it undermines the decision-theoretic strategy's applicability. If claims about branch number are strictly meaningless, then can the idealized betting scenarios appealed to in Wallace's argument, in which branch number is well defined, adequately represent real-life scenarios? Wallace is sensitive to this concern, and has recently attempted to address it (Wallace 2012: 183). He offers an extension of the argument which shows that, for a given quantum decision problem, the choice of coarse-graining of the consistent history space can be varied without altering whether a given preference order definable over acts satisfies all of the axioms. I shall assume that this response succeeds. My aim in this section is to argue that treating ascriptions of branch number as literally meaningless, or as never true, is not necessary to motivate branching indifference; I am happy to grant that it is sufficient.

To take stock: if we supplement EQM with the physicalist functional conception of preference and with the assumption of branching homogeneity, or if we reject talk about branch number altogether, then we can motivate branching indifference. However, both of these strategies are problematic. Instead I propose an alternative strategy drawing on Individualism.

According to Individualism, other Everett worlds comprise alternative possibilities. Introducing additional branching of Everett worlds, where each newly created Everett world contains the same reward as the parent Everett world, is then a matter of multiplication of alternative token possibilities; but the number of alternative possibility types is unaffected. Given that the actual Everett world is in some set of Everett worlds characterized by a particular reward—that is, given that a certain reward-specifying proposition is true at the actual world—we should be indifferent to how many other Everett worlds the salient set happens to contain. In other words, we should be indifferent to how many ways there are for some reward-specifying proposition to be true. What we care about is invariably whether that proposition is true at the actual Everett world: about whether the reward in question is actual.

Wallace himself, in earlier work, used an argument for branching indifference which is very similar to the one I have just given. There he presented the argument as depending not on Indexicalism, but on 'subjective uncertainty' (SU), the thesis that rational uncertainty about the result of an upcoming measurement is available for an Everettian agent:

> Suppose that someone proposes to increase a million-fold the number of the agent's descendants who see heads: say, by hiding within the measurement device a randomizer that generates and displays a number from one to 1 million, but whose output the agent doesn't care about and probably never sees. Then from the SU viewpoint, this just corresponds to introducing some completely irrelevant extra uncertainty. For it is the central premise of the SU viewpoint that [a] process which from an objective standpoint involves branching, may be described subjectively as simply one with uncertain outcomes. In this case the objective description is 'the agent branches into a million copies who see heads, and one copy who sees tails'; the correct description for the agent himself is 'I will either see heads or tails, and I am uncertain as to which; if I see heads then I am further uncertain about the result of the randomiser reading—but I don't care about that reading'. But it is a (trivially) provable result of decision theory that introducing 'irrelevant' uncertainty of this kind is indeed irrelevant (*it is essentially the statement that if we divide one possible outcome into equally-valuable suboutcomes, that division is not decision-theoretically relevant*). As such, from the SU viewpoint branching indifference follows trivially. Wallace (2006); emphasis added

This argument trades essentially on a treatment of branches as alternative possible outcomes. Without Individualism, the 'equally-valuable subout-comes' are not alternative possibilities, so it makes no sense to be uncertain about which of them will occur, and the argument lapses.

With Individualism on board, the population-ethics objection is defused. The analogy between a choice of quantum act and God's offer to create copies of the solar system rested on a tacit assumption of Collectivism. The copies God offers to create would, if created, be part of the actual world. Collectivism entails that people brought into existence by extra branching processes are likewise part of the actual world, and hence it seems that just as it is rationally permissible (and indeed, perhaps rationally required) to value the copies God creates, it is rationally permissible (and indeed, perhaps rationally required) to value the additional people that result from additional branching of Everett worlds.

Once we adopt the full package of Indexicalism, the disanalogy between a choice of quantum act and God's offer comes clearly into view. The people God offers to create would be actual, while those brought into existence by additional branching would be (according to Indexicalism) merely possible. Assuming that we ought rationally only to care about actual things, Indexicalism thus entails that we ought rationally not to care about the other Everett worlds and their inhabitants. Of course, even if Indexicalism is granted, branching indifference may still be rejected via a stubborn insistence that an act's (direct or indirect) consequences for actuality do not exhaust its normative significance. This line of resistance comprises what I have called the ethical objection to quantum modal realism; I responded to it in section 1.9.

I hope the shape of the Quantitative argument for Indexicalism is now clear. Indexicalism provides a compelling rationale for Wallace's principle of branching indifference, and so it plays a key role in solving the Quantitative problem for Everettian probability. Indexicalism thereby earns its keep as part of the best Everettian physics–metaphysics package deal.

In section 3.6, I present my third and final argument for Indexicalism: it plays a key role in solving the Epistemic problem for Everettian probability.

3.6 The Epistemic Argument for Indexicalism

The application of ordinary epistemological principles to the confirmation of EQM raises some well-known puzzles. In this section I shall focus on two potential puzzles that even the mere epistemic possibility of the truth of EQM raises for our usual understanding of how theories are confirmed by evidence. These are the *automatic confirmation puzzle* and the *quantum doomsday puzzle*.[15] According to the automatic confirmation puzzle, any experimental results whatsoever (as long as they are not directly inconsistent with EQM) provide support for EQM over a one-world stochastic theory. And according to the quantum doomsday puzzle, coming to accept EQM has radical epistemological consequences: Everettians ought to become virtually certain that the world is about to end. Crucially, these puzzles do not just arise for convinced Everettians; *anyone* who assigns non-zero prior credence to EQM must explain why our observations do not provide automatic confirmation of EQM over a one-world stochastic theory, and why we ought not to regard overlapping Everettian scenarios with utter dread.

The automatic confirmation puzzle and the quantum doomsday puzzle both raise serious problems, and their combination is especially unpalatable. Fortunately, the Indexicalist framework for Everettian probability that I have developed in this chapter allows us to avoid both of them. This encouraging result forms the basis for my third and final argument for Indexicalism: the Epistemic argument.

First, the puzzle of automatic confirmation. This puzzle arises when EQM is being compared with some candidate one-world stochastic approach to quantum theory[16]—call it ST—because EQM and ST have systematically different consequences for the number of observers in existence. Consider the following case (illustrated in Figure 3.2):

> *Quantum Wombat:* Wombat is unsure whether EQM or ST is correct. She has just performed a spin measurement with possible outcomes Up and Down, but she has not yet looked at the result. According to EQM, after the measurement there are two observers, located on branches of equal weight,[17] one of whom observes Up and the other of whom observes Down. According to ST, after the measurement there is one observer, who observes either Up or Down, with equal probability. Wombat is not sure whether *i)* EQM is true and she is one of the two observers, or *ii)* ST is true and she is the only observer.

[15] I have treated these two puzzles in detail in recent papers (A. Wilson 2014, A. Wilson 2017) and for the full story I refer the reader to those papers.

[16] I assume a stochastic theory here only for simplicity; a Bohmian one-world account of quantum theory would work equally well as a contrast case.

[17] For our purposes, only the relatively straightforward equally weighted case will be needed.

Figure 3.2. Quantum Wombat

Suppose Wombat observes Up. Should she take this observation to support EQM over ST? Plausibly not: many-world and one-world versions of quantum theory are usually supposed to be empirically equivalent. However, the probability that *Up is observed by somebody* is 1 according to EQM but is less than 1 according to ST. Accordingly, the credential update rule that Greaves (2007a) and Bradley (2011) dub 'naïve conditionalization' appears to break down when one of the options on the table is EQM.

Naive conditionalization endorses the following tempting line of thought: if the many-worlds theory predicts all possible outcomes, then no possible observation can disconfirm EQM; and since no one-world theory likewise predicts all possible outcomes,[18] every observation confirms EQM over ST. In Bradley's words: 'The Ancients could have worked out that they have overwhelming evidence for [EQM] merely by realizing it was a logical possibility and observing the weather' (Bradley 2011: 336). Something has gone wrong. And recall that it was not any assumption that EQM was correct that landed us in this mess; it was just the assumption that EQM might be correct. Compensating for the automatic confirmation by reducing our prior credences in EQM is an unattractively brute-force response,

[18] I set aside what Bostrom (2002) calls 'big-world cosmologies': single-world theories where a spatial or temporal infinity ensures that some plenitude of qualitative possibilities is realized.

especially since branching is so ubiquitous.[19] We can only safely ignore the apparent problem that EQM generates for confirmation theory if we are willing to set our prior credence in EQM arbitrarily low, effectively ruling out EQM a priori. This seems an uncomfortably dogmatic position for philosophers to adopt, given how seriously physicists take EQM. So the automatic confirmation puzzle is a puzzle for non-Everettians too.

Second, the quantum doomsday puzzle. Here, the source of perplexity is that a well-known feature of self-locating probability theory, the so-called *doomsday effect*, applies in a particularly savage way in the context of EQM. The doomsday effect is exploited in the *doomsday argument*, a curiosity of probability theory which has been independently identified several times over; the best-known sources are Carter (1983), Gott (1993), and Leslie (1989). Driving the doomsday argument is the idea that evidence about our *birth rank*—about the number of humans born prior to us—confirms the hypothesis that the total population of humans will not be very much greater than the total up to the present moment.[20] Informally: if doom lies in the distant future, then we are living unusually early in human history, whereas if doom will be within the next few hundred years, then we are not living unusually early.

The ordinary doomsday argument relies on a disparity between current population levels and past population levels.[21] Since current levels are so much higher than historical levels, if humanity goes extinct soon then most of the people ever to have lived will have lived only a few centuries before extinction. This disparity in population levels, however, is utterly dwarfed by the disparity in population levels between early and later times that is entailed by overlapping EQM, the 'splitting worlds' version of EQM considered and rejected in chapter 2.

[19] The issue of how, if at all, to quantify branching in EQM is a vexed one: see Wallace (2012) and Greaves (2007b). However, given that decoherence takes hold on timescales in the order of 10^{-20} seconds (Zurek 2002), any reasonable criterion would require initial prior credence in EQM to be well below 10^{-20} in order to insulate us from the automatic confirmation effect.

[20] It has sometimes been suggested (e.g. by Monton 2003) that the doomsday argument can be run without the need for knowledge of birth rank. In my view, Bradley (2005) convincingly argues that Monton's argument neglects observation selection effects.

[21] The version of the argument given by Gott (1993) makes no assumption about the distribution of humans throughout human history, and thus has less dramatic consequences then the version given by Leslie (1989), which takes into account the recent dramatic rise in population. (Sober (2002) refers to these arguments as 'Gott's line' and 'Leslie's wedge' respectively.) My argument here is analogous to Leslie's. See also Bradley and Fitelson (2003) for further distinctions amongst versions of the argument.

On the overlapping version of EQM, Everett worlds are constantly splitting into multiple copies of themselves. The corresponding re-fusion is suppressed by the temporally asymmetric process of decoherence, which produces branching of worlds towards the future but no branching towards the past.[22] This temporal asymmetry of decoherence leads to a disparity in the number of Everett worlds between early times and later times. Most world-stages are closer to the end of the Everett world of which they're a part than to its beginning. There is a corresponding disparity in the number of inhabitants of Everett worlds between early times and later times. Most people are closer to the end of the population of which they're a part than to its beginning.

The more branching, the larger the quantum doomsday shift. Since branch number grows exponentially, the size of the shift is gigantic. For example, if every branch splits in two once per second, the number of branches is 2^n after n seconds. If the branching is into more than two branches, the disparity becomes even more striking. In the bifurcating case, after n seconds the number of branch-temporal-parts in existence, 2^n, is greater than the total number to have existed up to that point (2^{n-1}). And if t_2 is 10 branching-events after t_1, then there are 2^{10} more branches at t_2 than there are at t_1. Moreover, branching is utterly ubiquitous. Decoherence occurs on extremely rapid timescales, and the branching is into countless branches. Even if branch number is in some sense or other indeterminate, it is determinately gigantic. So the quantum doomsday shift is overwhelmingly larger in size than the ordinary doomsday shift. As a result, it seems that rational Everettian agents should expect doomsday overwhelmingly soon. This is the quantum doomsday puzzle.

One straightforward way to avoid the quantum doomsday shift is to give up on overlapping EQM, and to adopt diverging EQM instead. According to diverging EQM, the number of Everett worlds does not increase over time, so there is simply no temporal asymmetry of branching of the kind that the quantum doomsday argument exploits. There are as many Everett worlds before a measurement interaction as there are after a measurement interaction; so, while the original doomsday argument stands, the quantum doomsday argument lapses. But the problem with adopting this easy way out of the quantum doomsday puzzle is that it makes an epistemic

[22] The ultimate source of the temporal asymmetry of decoherence presumably lies in a low-entanglement initial quantum state of the universe (Wallace forthcoming). This will not matter for present purposes.

question—the question of how EQM can get confirmed—depend on what looks like a purely metaphysical question—the question of how to interpret the decoherent histories formalism in the face of the apparent underdetermination highlighted in chapter 2. Consequently, it would be desirable to find a way to block the quantum doomsday effect independently of the question of divergence vs overlap. As I will argue, the principle Individualism which is incorporated into Indexicalism provides a way to do just that.

Indexicalism allows us to handle both the automatic confirmation puzzle and the quantum doomsday puzzle in a unified way. To see how this works, we need to return to the decision-theoretic treatments of Everettian probability discussed in the previous section. The most sophisticated decision-theoretic treatments—those due to Wallace (2006, 2010b, 2012), Greaves (2007a), Saunders (2010b) and Greaves and Myrvold (2010)—model both the epistemic state of an agent certain of the truth of EQM and of an agent wondering whether EQM is correct. These authors (henceforth: WGSM) fuse single-world scenarios and many-world scenarios into a single epistemology and describe a process of rational choice between theories positing such scenarios.[23] I will show how the WGSM decision-theoretic treatments resolve the automatic confirmation puzzle and the quantum doomsday puzzle, and then argue that these resolutions only succeed in the context of a background metaphysics that incorporates Individualism.

First, how do the WGSM decision-theoretic treatments handle the automatic confirmation problem? They do so by constructing decision theories that are neutral between EQM and arbitrary one-world stochastic theories. The first stage of the WGSM treatment is to model the situation of an agent wondering whether EQM is correct as an epistemic decision problem. On being presented with new evidence, we perform an epistemic 'act' of adopting a new credence function; we therefore need a decision theory to describe the choice between available epistemic acts. Crucially, for present purposes, this decision theory needs itself to be neutral between EQM and one-world theories.

Since we are considering an epistemic decision theory, the possible outcomes of an agent's possible epistemic acts consist in that agent's beliefs having varying levels of accuracy. If neutrality is to be achieved between one-

[23] These treatments differ in various ways. In particular, Wallace's project is more ambitious than that of Greaves or of Greaves & Myrvold, in that Wallace aims to prove an Everettian analogue of the Principal Principle connecting chance and credence; however, these differences will not matter for present purposes. Wallace's argument is explored in more detail in section 3.8.

world and many-world theories, it is crucial that accuracy be appropriately measured within the context of each theory. According to the WGSM treatments, the accuracy to be maximized in theory choice is accuracy with respect to the *decision-theoretic state*; and, according to the WSGM treatments, in Everettian contexts the role of the decision-theoretic state is to be played by an individual centred Everett world (an Everett world, plus an agential perspective within that world) rather than by an entire centred Everett multiverse.[24]

This needs a bit of unpacking. A decision-theoretic state[25] is ordinarily thought of as the state of an agent's actual environment; but what matters for epistemic-decision-theoretic purposes is that the accuracy of a rational agent's beliefs is assessed with respect to the state. In non-Everettian contexts, states are often taken to be centred possible worlds.[26] But in the Everettian scenario there are two potential candidates to play the role of a decision-theoretic state: the agent's centred Everett world and the agent's centred Everett multiverse. Greaves puts the point very clearly:

> The key question is what, in our framework... we should identify with Savage's 'States'. There seem to be two options:
>
> 1. For a decision taken at time t, take the set of States to be the set $W^C{}_t$ of centered worlds at time t.
> 2. Take the set of States to be the set H of histories.
>
> When there are no branching multiverses on the table, these two options coincide; so our existing usage of decision theory fails to commit us to one or the other... It is absolutely crucial to the Everettian program that option (2), not option (1), is the right extension. (2) is the option that leads to the strictly stronger decision theory, according to which branch weights can function, for all practical purposes, just like chances. If only option (1) is defensible, the Everett interpretation is sunk. Greaves (2007a)

The key question is: should we take observed statistical evidence to be evidence about which Everettian multiverse we're in, or should we take it to be evidence about which world we are in within an Everettian multiverse?

[24] This move is made, more or less explicitly, by Wallace (2012: 219–21), by Greaves (2007a: 128–9), and by Greaves & Myrvold (2010: 296–8). See also Greaves (2007a: 147).

[25] This terminology derives from Savage (1972).

[26] I shall work for convenience in a 'timeless' decision-theoretic framework. This simplifies the presentation and is adequate to epistemic decision scenarios—such as those relevant to the Doomsday Argument—which lack any significant diachronic element.

As Greaves argues, Everettians need to take the latter course if they are to avoid the automatic confirmation problem.

While Greaves is correct that her option (2) is necessary to avoid devastating confirmational consequences, she gives no independent reason to adopt it instead of option (1)—and, indeed, she acknowledges that the intuitive connection between chance and uncertainty puts strong pressure towards option (1). But she resists this intuitive pressure, recommending that Everettians adopt option (2) on the basis that it is 'an innocuous extension of claims that have been defended elsewhere in the literature'. Here Greaves is referring to the solutions to the Incoherence problem given by Greaves (2004) and by Wallace (2003b, 2007). The problem is that, even if (2) is an innocuous extension of these solutions, the solutions themselves are inadequate—as I have argued in section 3.4 and in more detail in Wilson (2013a).

Indexicalism provides an alternative motivation for option (2). In section 3.4 and in Wilson (2013a) I argued that the connection between uncertainty and objective probability that Greaves and Wallace opt to sever is a central aspect of the chance role, but that this connection can be restored by adopting Indexicalism. Since Indexicalism provides a well-motivated solution to the Incoherence problem, it allows for the WGSM decision theories to be applied to the Everettian scenario in such a way as to resolve the automatic confirmation puzzle. Without Indexicalism, the choice of Greaves's option (2) lacks any independent motivation. Indexicalism thus emerges as indispensable to a well-motivated Everettian solution to the automatic confirmation puzzle.

Next, how do the WGSM decision-theoretic treatments handle the quantum doomsday puzzle? Through their identification of decision-theoretic states with Everett worlds, the WSGM decision-theoretic treatments effectively factor uncertainty about self-location within a branching multiverse into two component uncertainties: self-locating uncertainty within a single Everett world, and uncertainty about which Everett world one is in. The treatments then impose the following constraint (which corresponds to the Born rule) on credences about which Everett world one is in: these credences, conditional on the truth of EQM, should match the weights assigned to each Everett world. However, they place no special constraints on self-locating uncertainty within a single Everett world. Consequently, the WSGM treatments leave open the possibility of applying something like Elga's restricted indifference principle (Elga 2000, 2004) within any specific Everett world, but they rule out the possibility of applying it across Everett worlds. While indifference reasoning turns on relative numbers of outcomes, branch weights are insensitive to fine-grainings of the history space, which means that altering the number of worlds on which an outcome occurs will not affect the probability of that outcome.[27]

In the WSGM decision-theoretic frameworks the indifference principles driving the quantum doomsday argument cannot be applied across different branches of the multiverse, so the argument for the quantum doomsday shift breaks down. Since the WSGM decision-theoretic treatments are neutral between divergence and overlap, this response to the quantum doomsday argument dissolves the concern that a purely metaphysical issue is settling epistemological questions. Meanwhile, the original doomsday argument is unaffected. That argument, transposed to the Everettian setting, involves only self-locating uncertainty within a single Everett world, so the indifference principle that it involves does not clash with the Born rule. Nothing in the WSGM decision-theoretic treatments undermines Elga-style indifference reasoning when it is confined only to in-branch applications.

When doing decision theory in the classical one-world setting, in order to adequately account for self-locating uncertainty we need to need to treat the question of where in the world we are differently from the question of what the world is like. Likewise, if EQM is correct, in order to adequately account for self-locating uncertainty we need to treat the question of which Everett world we're in differently from the question of where we are in that Everett world. The WGSM decision theories can achieve this independently of the question of overlap vs divergence, so long as we restrict indifference reasoning from applying across distinct Everett worlds. But it seems rather unsatisfying, and somewhat ad hoc, to simply stipulate such a restriction when setting out an Everettian decision theory. Do we have any independent reason to think that the resulting decision theory correctly models agents in an Everettian context?

My suggestion—as before—involves Indexicalism. Judicious modification to principles connecting the physics of EQM with the metaphysics of modality results in a metaphysical framework from which a WSGM-style decision theory follows naturally. According to the Individualism component of Indexicalism, each Everett world is a different metaphysically possible world. According to the Indexicality-of-Actuality component, the actual Everett world just is the world in which we are embedded. Putting

[27] This holds true as long as the relevant decoherence conditions are satisfied; see Wallace (2012: §4.3).

these consequences together, Indexicalism allows us to treat individual Everett worlds as decision-theoretic states while still maintaining the original conceptual connection between the decision-theoretic state and the agent's actual environment. Indexicalism thereby vindicates the WSGM decision-theoretic treatments.

Indexicalists can also retain the letter, if not the spirit, of Huw Price's principle that 'where goes ontology, there goes possible preference' (Price 2010: 380): they need only interpret the principle with an implicit restriction to the ontology of the actual world. For Everettians who adopt an indexical view of actuality, this kind of quantifier restriction is unmysterious and endemic in ordinary talk. Likewise, the indexical conception of actuality allows us to retain a related principle that 'where goes actuality, there goes the epistemic possibility of self-location'. If 'actuality' refers indexically to an agent's own Everett world, then this principle is validated by the WSGM decision-theoretic frameworks. Uncertainty about one's location within an Everett world is treated as uncertainty about where in the actual world one is located, whereas uncertainty about which Everett world one is in is treated as uncertainty about ordinary contingent matters of fact.

Once we understand Everettian probability along Indexicalist lines, ordinary confirmation theory delivers the result that quantum mechanics is confirmed by evidence in the usual way. Likewise, Elga's indifference principle still gives rise to the original doomsday effect, but it gives rise to no quantum doomsday effect.[28] If individual Everett worlds are possible worlds, then Elga's indifference principle, without any need for ad hoc modification, already tells an agent only about how to divide credences between various locations within a single Everett world. Hence, a quantum modal realist understanding of EQM makes it possible to solve the quantitative problem with Everettian probability without modifying a very plausible principle of classical self-locating epistemology.

In this section I have argued for Indexicalism on the basis that it helps resolve the Epistemic problem with probability in EQM, and have shown how this plays out in the specific contexts of the automatic confirmation problem and the quantum doomsday problem. This completes my positive case for an Indexicalist treatment of Everettian probability. In section 3.7, I flesh out the resulting picture by showing that Everettian branch weights,

[28] The indexical conception of actuality also does not affect more familiar applications of Elga's indifference principle: the principle still applies in the usual way to the Sleeping Beauty puzzle (Elga 2000) and to Elga's duplication puzzle (Elga 2004).

understood along Indexicalist lines, fulfil all the important criteria that philosophers have associated with the theoretical role of objective probability.

3.7 Chance Naturalized

The arguments of the last three sections illustrate that branch weights behave very much like chances with respect to their relationships to expectation, to action and to theory confirmation. Still, we might continue to wonder whether branch weights *really are* chances. To address this question in accordance with the methodology for metaphysics adopted in the introduction, in this section I consider the set of 'platitudes' that have been taken to characterize the theoretical role of chance and argue that Indexicalism can vindicate them. The Indexicalist picture of Everettian probabilities that has emerged from this chapter satisfies all of these platitudes, once the platitudes are formulated in an appropriately theory-neutral way.

Once we have a measure over any space, and can show that it obeys the axioms of probability, then there remains no further question that that is *a* probability measure, in the mathematical sense. But all sorts of measures count as probability measures in that sense: if we divide a year into measurable durations (say, into months), then proportion-of-year will count as a probability measure. For something to be a genuine probability measure, it had better be defined over a space of alternative possibilities. I argued in section 3.4 that Individualism renders the branch weights a genuine probability measure in this sense, while Collectivism does not. But even granting Individualism, not all measures over the space of Everett worlds get to count as objective probabilities. That a measure satisfies the probability axioms and is defined over a space of kinematically possible histories by itself tells us nothing about what kind of probability it is, or about whether knowledge of its value should be of any interest to a rational agent. Our question then becomes: what features of a probability measure qualify it as an objective probability measure?

Following the general functionalist approach to chance advocated by David Lewis (1980/1983), which aligns well with the Quinean methodology described in section 0.3 of the introduction, I maintain that to be chance is to fit into rational inferential and decision-making procedures in the right sort of way. 'The right sort of way' obviously needs unpacking. The functional conception of chance has been factored by Papineau (1996) into the 'decision-theoretic link' (perform whichever action maximizes chance-weighted expected utility) and the 'inferential link' (use observed relative frequencies to guide your beliefs about the chances). Similarly, Saunders (2010b) proposes the following three principles as 'the most important of the chance roles':

> (i) Chance is measured by statistics, and perhaps, among observable quantities, only statistics, but only with high chance.
> (ii) Chance is quantitatively linked to subjective degrees of belief, or credences: all else being equal, one who believes the chance of E is p will set his credence in E equal to p (the so-called 'principal principle').
> (iii) Chance involves uncertainty; chance events, prior to their occurrence, are uncertain. Saunders (2010b)

Saunders's (i) and (ii) correspond respectively to Papineau's 'inferential link' and 'decision-theoretic link'; (iii) is an additional ingredient.

Disagreement amongst proponents of the functional conception should come as no great surprise; the functional conception, broadly construed, is silent on what precise ingredients make up the functional role of chance. Those such as Saunders who are inclined to seek conciliation between the everyday and the scientific worldviews typically build more conceptual content into the chance role than more sceptically inclined authors, such as Papineau and Wallace, who resist such extra content. Papineau, for instance, rejects both Saunders's condition (iii) and the related thought that chances are probabilities for some outcome to be actual:

> Normally we think that just one of a set of chancy outcomes will occur, with the probabilities therefore indicating the outcomes' differing prospects of becoming actual. On the [Everettian] view, by contrast, all the outcomes will definitely occur, on some branch of reality, and the probabilities therefore need to be read as attaching weights to these different branches. But it seems to me that this contrast is a 'dangler', which makes no difference to the rest of our thinking about probability.
>
> Papineau (1996: 239)

Instead of being regarded as excessive, Saunders's three principles governing the chance role might be thought insufficient. As well as accommodating 'operational' and 'common-sense' principles governing chance, those of a more metaphysical inclination have suggested that a satisfactory chance-candidate should accommodate a number of more explicitly metaphysical principles. For example, Schaffer (2007) discerns no less than six distinct ingredients in the chance role. His rough statements of these ingredients are as follows:

Credence: 'information about the objective chance of p should fix your credence in p onto its chance value' (ibid.: 123)
Possibility: 'if there is a nonzero chance of p, this should entail that p is possible, and indeed that p is compossible with the circumstances' (ibid.: 124)
Futurity: 'if the chance of p at [t] is between 0 and 1, then p must concern t's future.' (ibid.: 125)
Intrinsicness: 'chance values should remain constant across intrinsically duplicate trials' (ibid.: 125)
Lawhood: 'chance values should fit with the values projected by the laws of nature... via history-to-chance conditionals' (ibid.: 126)
Causation: 'chances should live within the causal transitions they impact' (ibid.: 126)

The Credence ingredient, which Schaffer takes to involve something along the lines of the Principal Principle of Lewis (1980/1983), corresponds to Papineau's decision-theoretic link (given the standard connection between credences and expected utilities), and to Saunders's principle (ii) (reading 'all else being equal' as 'for a rational agent in possession of no inadmissible information'). All are agreed that some kind of link is needed between objective probability and rational credence. It is less clear how Schaffer's other ingredients relate to Papineau's inferential link, and to Saunders's principle (i). This is a nice question in the metaphysics of chance, but I will not pursue it any further here. My primary concern is to argue that, once we adopt Indexicalism, the Everettian weights do play the full chance role. As such, it does not matter if there is some redundancy in the constraints I adopt. I will therefore provisionally treat Papineau's inferential link and Saunders's principle (i) as jointly pointing to a further constraint on the chance role, over and above Schaffer's six ingredients, which I call *Measurement*:

Measurement: Chances can be empirically discovered, but only in the long run through statistical inference, and only with high chance.

Likewise, if some of Schaffer's own conditions are judged to be otiose (perhaps because it is thought that a lack of operational significance renders them 'danglers', in Papineau's sense), then I will just have shown more than I needed to. Indeed, it looks as though four of Schaffer's ingredients—Possibility, Futurity, Intrinsicness and Causation—have no role to play in Papineau's rather bare version of the functional conception of chance. However, Saunders's condition (iii) does appear to match up reasonably well with Schaffer's 'Possibility' ingredient, if we assume that the kind of uncertainty involved is objective uncertainty.

Wallace (2006, 2010b, 2012) argues for a strongly minimalist conception of the chance role. According to Wallace, objective chance—or at least: '"the *scientific conception of objective chance*"' (Wallace 2006; emphasis in original)—is exhausted by the Principal Principle. Insofar as Wallace is explicitly abjuring extra metaphysical requirements on objective chance as unscientific, it might look as though he and Schaffer are merely talking past one another, and are discussing different concepts. Schaffer has an alternative and much richer concept of chance in mind as the target of his analysis. Another reading of the situation is possible, however: we could take Wallace as suggesting that any quantity which genuinely satisfies the Principal Principle will *ipso facto* satisfy all and any other (legitimate) requirements on the objective chance role. Indeed, Wallace (2006) does sketch an argument from the claim that some quantity satisfies the Principal Principle to the claim that we are justified in taking observed frequencies to measure that quantity:

> For suppose we do have a theory T which allows us to define some magnitude C, and suppose PP_C is the proposition that C satisfies the functional definition (we might loosely say: satisfies the Principal Principle — hence the notation). It follows that if we accept both T and PP_C, we should set our credence in an event E equal to C(E). If C(E) is high and our prior credence in E is much less high, we should regard (T&PP_C) as explanatory of E, and thus regard E as reason to accept T and PP_C.
>
> Wallace (2006)

Set aside the question of whether this argument succeeds. Whether it does or not, it is not the only possible way of arguing that observed statistics measure some quantity which is a candidate for the chance role. We will encounter in section 3.8 a different way of arguing for this type of conclusion, one rather more specific to EQM. This highlights the possibility that certain specific constraints on the chance role might be philosophically interesting in their own right, even if satisfaction of them by some chance-candidate C is already entailed by C's satisfaction of the Principal Principle. It may matter *why* the constraints hold.

The point bears emphasis. Perhaps a version of the Principal Principle can be proved for some chance-candidate, and perhaps this will entail that the candidate satisfies all of the other constraints on the chance role. But if a proof of the Principal Principle turns out to be unavailable, or if satisfaction of the Principal Principle fails to entail satisfaction of the other elements of the chance role, then we might still be able to argue directly that various other elements of the role are satisfied by some chance-candidate. Perhaps we will never be in a position to prove, for some chance-candidate C, that C governs rational credence according to the Principal Principle. (For more discussion of this possibility, see section 3.8.) We might nonetheless be in a position to prove that C satisfies various constraints on the chance role which themselves are entailed by the Principal Principle. For this reason, it will be helpful to continue to keep all of the various constraints we have so far encountered on the table. If satisfaction of some of them entails satisfaction of some of the others, so be it.

As it happens, there seems to be good reason to think that at least one of these conditions does not follow from the Principal Principle; viz., the Intrinsicness requirement. Arntzenius & Hall (2003) have argued that certain functions satisfy the Principal Principle perfectly, but do not satisfy the Intrinsicness constraint (see also Schaffer 2003). Similarly, the 'omniscient' probability function which assigns probability 1 to all truths and probability 0 to all falsehoods satisfies the Principal Principle. Since these functions are not plausible candidates for the chance role, they are counterexamples to the claim, made by both Wallace (2006) and Lewis (1980/1983), that the Principal Principle captures everything we know about chance.

Putting all this together, we end up with eight mooted constraints on functions which are candidates to play the chance role (some possibly overlapping, some possibly dispensable):

1. *Chance-formal* Chance is a probability function from propositions, worlds and times to the open unit interval: $ch(p,w,t) \rightarrow [0, 1]$.[29]
2. *Futurity* If $0 < ch(p_e,w,t) < 1$, then $t < t_e$. (p_e says that event e occurs; t_e is the time of e.)

[29] I assume with Schaffer that only worlds, times, and propositions are inputs to the chance function. Authors including Glynn (2010) and Handfield & Wilson (2014) have argued that we should in addition relativize chances to levels of theoretical description, but we can neglect that prospect for present purposes.

3. *Intrinsicness* If e' is an intrinsic duplicate of e, and the mereological sum of the events at t' in w' is an intrinsic duplicate of the mereological sum of the events at t in w, then $\text{ch}(p_e,w,t) = \text{ch}(p_e',w',t')$.
4. *Causation* If $\text{ch}(p_e,w,t)$ plays a role in the causal relation between c and d, then $t_e \in [t_c, t_d]$.
5. *Lawhood* If $\text{ch}(p,w,t) = x$, then the laws of w entail a history-to-chance conditional of the form: if the occurrent history of w through t is H, then $\text{ch}(p,w,t) = x$.
6. *Measurement* There is a high chance that observed statistics of outcomes of chance processes provide a reliable guide to the chances.
7. *Possibility* If $\text{ch}(p,w,t) > 0$, then there exists a world w' such that: (i) p is true at w', (ii) w' matches w in occurrent history up to t, (iii) w' matches w in laws.
8. *Credence* Evidence that $\text{ch}(p,w,t) = x$ provides a reason in w to have credence x in p. (This condition will be precisified in what follows.)

My aim in the rest of this section will be to argue that Everettian weights meet all eight of these conditions on the chance role.

Chance-formal must be addressed before the other constraints, since the assumptions required to ensure that it holds for Everettian weights will help us to properly assess whether weights can meet the remaining conditions. It is entirely uncontroversial that the weights satisfy the formal conditions required of a probability measure—that is, they form a measure over a space such that the measure of the entire space is equal to 1. But Chance-formal requires more than that; it also requires that the inputs to the probability function be propositions, worlds, and times.

Individualism ensures that weights are defined over a space of possible histories, and hence captures the 'world' input to the chance function. But this by itself is insufficient to capture all of the content of Chance-formal, since probability measures definable over spaces of possible histories are not necessarily functions of propositions and times. The addition of the other components of Indexicalism vindicates Chance-formal in full, by connecting propositions, worlds and times to the decoherent histories formalism in the right way. Specifying an Everett world w and a time t picks out a segment S of w—the history of w up to t. Multiple distinct worlds will have initial segments which are duplicates of S—call the set of all such worlds *Dset*. The weights will form a measure over *Dset*. According to Indexicalism, specifying a proposition P picks out a set of worlds—call this set *Pset*. The intersection of *Dset* with *Pset* will be a further set of worlds—call it *Iset*—that contains all of the worlds that have the same initial segment as w up to t, and at which P is true. If we sum the weights of the worlds which are members of *Iset*, then we obtain the total weight of worlds which duplicate w up to t and at which P is true. Dividing this quantity by the total weight of the worlds in *Dset* provides a number in the open unit interval, as required; and this number is a function of worlds, times and propositions. Indexicalism thus entails Chance-formal for Everettian weights.

Note that the argument just given required us to treat Everett worlds as alternative possible worlds; that is, it presupposed Individualism. This conclusion was anticipated in section 3.3, where I argued that only Individualism allows us to solve the incoherence problem and explain what objective probabilities in EQM are probabilities *of*. Chance-formal is just a rigorous statement of the requirement that the proposed chance-candidates should be structurally the right sort of things to count as chances; it is therefore no surprise that Individualism is required if Chance-formal is to be satisfied in EQM.

I shall discuss the remaining conditions in order of how easy they are to satisfy. First I shall argue that the Futurity, Intrinsicness, Causation, Lawhood, and Measurement conditions are straightforwardly satisfied by Everettian weights; only the basic properties of EQM are needed to establish these results. However, I shall also argue that traditional versions of EQM do not permit the weights to satisfy Possibility, and that Indexicalism is needed here. Credence is the most demanding constraint. One of the most promising approaches to arguing directly for the principle lies in the decision-theoretic strategy of Deutsch and Wallace; I have already argued in chapter 3.5 that this strategy relies on the ingredients of Indexicalism. In section 3.8 I discuss the prospects and limits of this strategy, and assess the extent to which it vindicates Credence.

I shall illustrate the holding of various conditions using a toy 'chance set-up': a spin measurement with two possible outcomes, Up and Down. Applying the Indexicalist picture, this chance set-up may be modelled by two sets of Everett worlds, such that in both sets the pre-measurement segments of each world are all qualitative duplicates of one another. Every world in one of the sets is an Up-world; every world in the other set is a Down-world. Let the total weight of the Up-worlds be equal to the total weight of the Down-worlds. Let p be the proposition that the outcome will be Down, as uttered at time t by an experimenter in one particular Up-world, w; p is therefore false at w.

First, the Intrinsicness condition. EQM itself guarantees that Intrinsicness will be satisfied. Weights are an integral part of the dynamical structure of quantum mechanics, and they supervene on the intrinsic properties of quantum systems. As a result, if we identify chances with weights, intrinsic duplicate systems will not differ with respect to the chances of any outcomes. Assuming EQM, weights therefore satisfy the Intrinsicness condition.

Consider next the Futurity condition. This also is straightforwardly satisfied by weights in EQM. Futurity requires that if the chance of event *E* occurring is between zero and one at time *t* in world *w*, *E* must be in the future at *t*. In terms of the Indexicalist picture, this requirement is that if some but not all worlds which duplicate *w* up to *t* are *E*-worlds, then *E* is not in the segment of *w* up to *t*. Weights accordingly satisfy the Futurity condition if worlds diverge but do not reconverge. This asymmetric structure is built into contemporary EQM through the role of decoherence, and may ultimately be traced back to a low-entanglement boundary condition on the early universe; for further discussion, see Wallace (forthcoming). Assuming EQM, weights therefore satisfy the Futurity condition.

The Causation condition requires that if a given chance is to explain the transition from cause to effect, then that chance must concern some event within the time interval from when the cause occurs, to when the effect occurs. Let us change the toy example slightly, so that the Up-worlds have combined weight 0.99 and the Down-worlds have combined weight 0.01. Presumably this is a case where the high chance of Up is a candidate explanation for the falsity of *p*. The causal transition in question is a transition from the measurement apparatus being triggered to a result of Up being obtained. Since the point of divergence between the Up-worlds and Down-worlds occurs during this transition, then the weights satisfy the Causation condition.

The Lawhood condition is that the laws of a world, together with an initial segment of that world up to time *t*, should determine the chances at time *t*. In the context of weights, this requirement is that the laws of quantum mechanics, together with some segment of a world up to time *t*, should determine the weights at *t*. This is validated by EQM; since weights are themselves dynamical quantities, physical attributes of the worlds, specifying a world segment *S* up to *t* is enough to fix the relative weights of all worlds whose initial segments duplicate *S* up to *t*.

Satisfaction of the Measurement condition also follows from the principles of EQM, without any need to bring in the framework of Indexicalism. Arguments of this general form trace back to Everett's dissertation (Everett 1957a), where it was shown that the weights have the mathematical form of a measure over histories, and an analogue of Bernouilli's theorem (also known as the 'law of large numbers') was proved. The result is that worlds where long-term observed statistics diverge from the predictions of the Born

Rule have a much lower combined amplitude than worlds where statistics are in approximate agreement with the Born rule. Saunders (2010b) provides a purely dynamical proof of a similar result; that for a set of duplicate quantum systems subjected to a measurement, the measured relative frequencies of each outcome are close to the weights *in worlds which themselves have high weight*. The proof relies only on the properties of the unitary dynamics and on the exchangeability (in the sense of de Finetti 1931) of the measurement protocol.

In addressing constraints 2–6 I have not needed to rely on any elements of the Indexicalism picture, except for ease of exposition. I appealed only to the structure of unitary quantum mechanics, which is common to all versions of EQM, including fission programme versions which combine overlap of worlds with Collectivism. However, non-Indexicalist approaches run into difficulties with the Possibility condition, which requires that if there is a non-zero chance of P at t in w, then P is compossible with w's past history at t. Given Indexicalism, this amounts to the requirement that if some worlds whose initial segments duplicate w up to t are P-worlds, then P was a genuine possibility at t in w. This follows from Indexicalism, which entails that for the experimenter in w at t the other Up-worlds and the Down-worlds comprise alternative future possibilities, which cannot be ruled out by any evidence available at t. Assuming Indexicalism, weights therefore satisfy the Possibility condition.

3.8 An Everettian Principal Principle

The last of our conditions is Credence. While the justification of chance-credence links is controversial given all conceptions of chance, such links are a core part of the chance role. Whether or not we agree with Lewis that the chance-credence link captures 'everything we know about chance' (Lewis 1980/1983), the dependency of rational credence on knowledge of the objective chances is common ground between all adherents of the functional conception of chance.

The Credence condition can be precisified in various ways. The best-known version is David Lewis's Principal Principle:

Principal Principle: For any rational initial credence function C, for any proposition p, for any evidence E admissible[30] at time t, and where X is

[30] Admissible evidence, roughly, is evidence which does not pre-judge the outcome of any chance events. See Meacham (2005) and Hoefer (2007) for illuminating discussions.

the proposition that the objective chance of p at t is x: $C(p/XE) = x$. (Lewis 1980/1983)

Given the quantum modal realist theoretical identification of objective chance, the Lewisian principle can be retained unchanged in the Everettian context. But to help see more transparently how this goes, we can formulate the following EQM-specific version of the principle:

Everettian Principal Principle (EPP): For any rational initial credence function C, for any proposition p, for any evidence E admissible at time t, and where X is the proposition that the proportion by weight of Everett worlds matching the actual Everett world up to t at which p is true is x: $C(p/XE) = x$.

To complete the argument that Everettian weights can play the chance role, we need a direct argument for the EPP; or at least, we need some explanation of how Everettians can legitimately adopt the EPP without giving a direct argument for it. It is time to turn to the recent work of David Wallace (Wallace 2010b, 2012), which is the most well-developed and systematic response to these challenges.

Wallace's argument is complex, and I will not reconstruct all of the technical details. Rather, I shall work with a subset of the premises Wallace uses and provide an informal argument from them for the Everettian Principal Principle. I think this informal reasoning is at the conceptual core of Wallace's argument; however, in Wallace's full proof the informal argument is intertwined with a representation theorem that recasts the foundations of probability from scratch in an Everettian setting and incorporates a Bayesian framework for theory confirmation. My aim is to set these complexities aside and use the informal argument to explore the limits of what can and cannot in principle be proved from Wallace's axioms. My conclusion will be that while Wallace cannot prove the Everettian Principal Principle from non-probabilistic premises (an achievement which has seemed like pulling a rabbit from a hat to many readers) he can nonetheless show that quantum branch weights are the uniquely best candidates for playing the Credence part of the chance role in the Everettian setting. We can then at least draw a conditional conclusion: if there are chances at all in the setting of EQM, then they are the quantum branch weights.

The premises of Wallace's to which I will be appealing are *State Supervenience* and *Branching Indifference*. I will also be making some assumptions about the nature of chance, and about the nature of rational value. I will assume that chances are rationally relevant to agents' preferences over acts, but that they are not something that can be directly valued by agents: agents care about chances only because they care about the outcomes to which chances attach. I will also be assuming that rational agents can value only objective physical quantities that it is in principle physically possible for them to detect: if a quantity is not something that can make any causal difference to the agent, then it makes no sense for an agent to assign value to it. The informal argument from these premises runs as follows:

1. Only objective physical features of the Everett worlds diverging from the actual Everett world only in the future are relevant to the preferences of a rational agent who knows that EQM is correct.[31] (State Supervenience)
2. Given EQM, (absolute or relative) branch weights are the only objective physical features of some Everett worlds diverging at some time t not physically possibly detectable by agents in those worlds prior to t. (Premise)
3. A rational agent can directly value any feature of reality if and only if it is physically possibly detectable by them. (Premise)
4. A rational agent who knows that EQM is correct can directly value any objective physical features of the Everett worlds diverging from their own only in the future except for the (absolute or relative) branch weights. (From 2, 3)
5. Chances are relevant to the preferences of a rational agent but are not themselves something that a rational agent could directly value. (Premise)
6. If there are chances in EQM, then they are (absolute or relative) branch weights. (From 1, 4, 5)
7. A rational agent who knows that EQM is correct is indifferent to the number of Everett worlds with a given outcome, as long as the total weight of Everett worlds with that outcome is held fixed. (Branching Indifference)
8. Rational agents are indifferent to absolute branch weights. (From 7)
9. If there are chances in EQM, they are relative branch weights. (From 5, 6, 8)

[31] Jansson (2016) argues that State Supervenience cannot be given a non-circular justification within the Everettian approach. I agree—see the last part of this section for discussion—but this does not undermine the interest of the argument to follow.

In the remainder of this section, I shall explain the two of Wallace's premises that I employ, and explain why I think this informal reasoning captures the conceptual core of his technical argument. I then go on to argue that the weaker conditional conclusion of the informal reasoning is all that Wallace could in principle succeed in establishing. I have already argued (in section 3.4) that one of Wallace's axioms, Branching Indifference, is only plausible if we have Individualism in the background. I have also argued (in section 3.3) that Indexicalism is needed to vindicate the connection between objective probability and objective uncertainty that is expressed in Saunders's condition (iii), and which is subsumed under Schaffer's Possibility constraint. Since I am aiming at an account of objective probability per se in EQM, rather than at an account of some surrogate for objective probability, I will freely assume the full Indexicalism picture.

State Supervenience, in an Indexicalist picture, says that a rational agent's preferences between acts depend only on the physical states of the worlds diverging after that act. Wallace defends this premise by noting that the rewards available to any agent depend only on the physical state of the worlds after divergence; and how some transformation affects states that are not the state of any of the worlds prior to divergence is irrelevant to the agent, since the agent is not in that situation. This justification for state supervenience goes through with or without Indexicalism; it requires only a weak and widely accepted form of physicalism.

Branching indifference, in an Indexicalist picture, states that an agent does not care about the number of worlds with a particular reward, as long as the combined weight of worlds with that reward is held fixed. In section 3.4 I described Wallace's dual defence of this axiom; the first component denies that there is any such thing as branch number, and the second component turns on the claim that branching (in the Indexicalist picture, divergence) is 'uncontrollable and ever-present' in EQM. I argued that Wallace's defence of branching indifference is inadequate as it stands, but adopting Indexicalism provides significant extra support for the axiom.

With these two premises granted, the argument proceeds by exclusion of objective physical features of reality other than the branch weights as potential candidates for being chances. Chances must make a difference to rational credence: that is part of what it is to be chance. But, by the broad form of physicalism embodied in State Supervenience, only objective physical features of the set of Everett worlds diverging from the actual world only in the future can make a difference to rational credence. Then, all objective physical features of that set of worlds other than the branch weights are excluded as potential candidates, since they are the sort of thing to which one could in principle rationally attach an intrinsic value. But chances, being measurable only retrospectively through frequencies, are not something to which one could in principle rationally attach an intrinsic value: they are valuable only insofar as they make valuable events more or less likely. Chances, on this picture, are real physical properties but they are properties *of* Everett worlds and not properties *within* Everett worlds.

Let me sum up the conceptual core of the argument. Chances are only of value to us insofar as the events of which they are chances are of value to us, but chances are nevertheless relevant to rational credence. In the Everettian setting, branch weights are the only objective physical features of reality that have the right profile to be chances; so branch weights *are* chances, if anything is. Weights are identified as chances in EQM by excluding all other viable candidates.

Wallace's argument incorporates the exclusion reasoning described above by invoking the need for an agent's preferences to respect the symmetries of a symmetrically weighted quantum state; but the exclusion reasoning is intertwined in his presentation with a representation theorem from the decision-theoretic tradition, and with a Bayesian approach to theory confirmation. While I have nothing but admiration for the robust epistemological framework that Wallace develops on behalf of Everettians, I think that the technical virtuosity with which Wallace presents this framework risks obscuring the status of the EPP. I alluded earlier to the apparently magical quality of the trick that Wallace claims to perform in deriving conclusions about rational credence from symmetries and other physical properties of the quantum state. I think this sense of conjuring derives from a certain overreach in Wallace's conclusions: where the informal argument allows only the conditional conclusion that in Everettian scenarios branch weight is chance if anything is, Wallace claims to have secured the stronger unconditional conclusion that branch weight is chance. Since this cannot be established by any argument whatsoever, something must have gone wrong somewhere.

It has become orthodoxy in the literature on the foundations of probability that the chance-credence link cannot be justified in a non-question-begging manner. Chances do not converge to frequencies by necessity, but only with a chance that approaches 1; so any demonstration that chances and frequencies converge must itself presuppose a chance-credence link. Apparent proofs of the link are usually diagnosed as tacitly relying on it in some way, as for example when Braithwaite (1957) criticized de Finetti's (1931) argument as tacitly relying on a chance-credence link when justifying the assumption of exchangeability. Strevens (1999) makes a general argument that chance-credence links cannot be justified for exactly the same reason that the problem of induction cannot be solved: there is no way of excluding the epistemic possibility that frequencies may radically diverge from chances indefinitely far into the future. I think Everettians should take this lesson to heart, and give up Wallace's goal of arguing that agents in Everett worlds who fail to set their credences by the Born rule are being *irrational.* Nonetheless, we can still make a strong case that if any chances exist in the Everettian setting then they are branch weights. If any chance-credence link is appropriate to EQM, then it is the EPP. That is enough, I think, to vindicate the Everettian theory of chance: it remains on at least as good a footing as any extant one-world theory of chance.

The case for parity between Everettian probability and orthodox one-world chance theories has been made several times in the literature, for example by Saunders (1998) and by Papineau (1996, 2010). These authors emphasize that we have nothing like consensus over any justification for the Principal Principle in any one-world metaphysical framework, and indeed no realistic expectation of discovering any justification. They also argue that Everettian weights have all of the necessary formal and functional features of an objective probability measure, features which in one-world stochastic physical theories are generally stipulated without argument. There is no justification for the Principal Principle in the context of the GRW theory, or Bohmian mechanics, or statistical mechanics. We should not hold it against EQM that it cannot do the epistemologically impossible. We may, if we wish, add that it is something like a Moorean truth, or intuitively obvious, that there are chances, or that chances are a precondition of the possibility of our epistemic practices. However, these considerations come from outside EQM proper.

While Wallace's decision-theoretic representation theorems do establish that a coherent epistemological and decision-theoretic framework is available to model Everettian agents, the theorems do not and cannot make any corresponding patterns of credences rationally required of those agents. Practical rationality simply does not constrain epistemic rationality in the way Wallace would require. An Everettian agent who has the preferences over actions that Wallace recommends but remains stubbornly certain that they are in a branch where they will win the lottery every week is not violating any principles of rationality. But we can solve the Everettian probability problem without needing to show that such an agent is epistemically irrational, so long as we can rest happy with the conditional identification of branch weight as chance *if anything is.* A conditional identification of this sort is all we have for any one-world indeterministic theory, and it is all we can reasonably expect from any physical theory. Wallace's constructions then stand as a demonstration of the possibility and consistency of an Everettian-friendly decision theory and confirmation theory; but he does not (and cannot) demonstrate that it is rationally obligatory for Everettian agents to adopt them. Such a demonstration would amount to a solution to Hume's problem of induction.

In this chapter I have focused on the well-developed decision-theoretic arguments for the Born rule due to Wallace and others. An alternative approach that has recently attracted attention is due to Sebens & Carroll (2018), who derive the Born rule from an epistemic principle they take to be a natural generalization of ordinary indifference reasoning to the cosmological context. While their approach raises many interesting questions, I think it remains subject to the same basic limitation as Wallace's approach; the indifference principle that is their key premise lacks non-circular justification, and so the argument cannot show that there is an unconditional rational requirement to treat weights as objective probabilities.

3.9 Summary

In this chapter I have shown how quantum modal realist principles connecting the physics of EQM with the metaphysics of modality provide a framework which is uniquely hospitable to a proper Everettian treatment of probability. With Indexicalism in hand, Everettians are in a strong position to respond to the Incoherence problem, the Quantitative problem, and the Epistemic problem, without the need to treat chance as an illusion or to revise highly plausible platitudes characterizing the chance role. From the Indexicalist perspective, a strong case can be made that Everettian branch weights satisfy—if anything does—the key constraints on objective chance that have been formulated in the literature. Branch weights plausibly satisfy all of the formal and functional criteria that make up the chance role. Accordingly, branch weights are the unique best candidates to play that role in Everettian scenarios. The principles which make up Indexicalism thus earn their keep in the service of a powerful and non-revisionary theory of Everettian chance.

The arguments for this conclusion have relied at several points on the novel metaphysical components of Indexicalism, echoing the main theme of this, and the previous, chapter: an adequate theory of Everettian probability relies on an adequate Everettian metaphysics of modality.[32]

[32] Some material from this chapter has been reproduced from my previous articles in *The British Journal for the Philosophy of Science*: 'Objective Probability in Everettian Quantum Mechanics' 64(4), December 2013, 709–37; 'Everettian Confirmation and Sleeping Beauty' 65 (3), September 2014, 573–98; 'The Quantum Doomsday Argument' 68(2), June 2017, 597–615.

4
Laws of Nature

4.1 Introduction

Quantum modal realists can give a unique and multifaceted theory of laws of nature. At root there are what I will call the *Fundamental laws of the multiverse*, the most general principles that characterize the fundamental physical reality underlying the emergent Everett multiverse. There are also what I will call the *fundamental laws of Everett worlds*, the most general principles which characterize the contents of all of the emergent Everett worlds. To distinguish these two kinds of laws in shorthand, I will write the former with an upper-case F and the latter with a lower-case f. The big-F Fundamental laws are deterministic and the small-f fundamental laws are indeterministic; both sets of laws are non-contingent. The Fundamental laws are analogous to the principle of recombination within Lewisian modal realism; the fundamental laws are analogous to the fundamental laws of individual Lewis worlds.

As well as the fundamental laws of Everett worlds, the quantum modal realist picture incorporates non-fundamental laws of Everett worlds of various kinds. These non-fundamental laws are typically restricted to apply to specific types of subsystems. An especially important subset of these 'local laws' are contingent, in virtue of incorporating parameters the value of which varies across different large-scale regions of the multiverse or of a single Everett world. The contingency of these non-fundamental local laws helps to reduce the degree of theoretical revision required of quantum modal realists in light of the non-contingency of the Fundamental and fundamental laws. Standard philosophical accounts of laws of nature treat all laws as contingent; but quantum modal realists may point to contingent local laws as a partial vindication of this familiar thought even in the presence of deeper non-contingent laws.

I begin in section 4.2 by setting the Everettian theory of laws of nature in the context of the vexed recent debate between *Humean* and *anti-Humean* theories of laws. Both of these approaches face serious problems, and debate

The Nature of Contingency: Quantum Physics as Modal Realism. Alastair Wilson, Oxford University Press (2020).

DOI: 10.1093/oso/9780198846215.001.0001

between them has effectively stalled; I describe the impasse, and suggest that a middle way is needed. I examine (section 4.3) one under-appreciated attempt, due to Robert Pargetter, to steer between Humeanism and anti-Humeanism, but I argue that this attempt founders on problems with the new primitive ontology that it introduces. Quantum modal realism can support a theory of laws that resembles Pargetter's but avoids the associated problems (section 4.4). I extend this theory of laws to account for special-science laws (section 4.5) and for laws that are localized to particular regions (section 4.6), and I use the latter form of laws to account for the widespread view that laws of nature are metaphysically contingent (section 4.7). A contextualist semantics for claims of lawhood is sketched (4.8), which sits naturally with a similar semantics for claims of objective chance. After discussing (4.9) how the ideology of naturalness of properties fits into a quantum modal realist picture, I assess (4.10) the performance of the quantum modal realist theory of laws against the desiderata identified previously, before concluding (4.11).

4.2 Humeanism and Anti-Humeanism

The most familiar dividing line in the debate over the metaphysics of laws of nature is between Humean and anti-Humean views of laws. These views differ primarily in what they say about the order of explanation between regularities in the occurrent matters of fact, on the one hand, and laws of nature, on the other. I'll express these disagreements about order of metaphysical explanation in terms of the idiom of grounding, conceived of as a worldly correlate of metaphysical explanation: grounding stands to metaphysical explanation as causation stands to causal explanation. As elsewhere in this book, if you distrust the notion of ground you can replace it with some more innocuous form of dependence that still entails (at minimum) supervenience.

According to *Humeanism about laws*, the laws of nature are fully grounded by occurrent matters of fact. It is just because the occurrent matters of fact are as they are that the laws of nature are as they are. *Anti-Humeanism about laws* is simply the denial of Humeanism: the laws of nature are not fully grounded in the occurrent matters of fact. Anti-Humeanism about laws leaves open both the question of how the laws of nature are in fact grounded, and the question of how the occurrent matters of fact may be explained by the laws.

The best-developed implementation of Humeanism about laws—David Lewis's Humean Supervenience (Lewis 1986b)—incorporates significant additional commitments over and above the core Humean claims. Humean Supervenience blends Humeanism about laws with a 'pointilliste' doctrine about occurrent matters of fact—that they comprise an assignment of intrinsic perfectly natural properties to spacetime points—and with an unrestricted principle of recombination for perfectly natural properties. However, other Humean doctrines about occurrent matters of fact are available; see e.g. Butterfield (2006), Schaffer (2010).

Humeanism about laws is supported by some strong arguments, but also faces serious difficulties. In its favour are epistemological, metaphysical and conceptual considerations:

Epistemological motivation: Humean laws are in principle knowable. Because we are in a position to have knowledge about occurrent matters of fact through direct observation, we are in a position to know about laws that depend on those occurrent matters of fact.

Metaphysical motivation: Humean laws are theoretically conservative. They are nothing other than complex redescriptions of occurrent matters of fact; if we are already committed to the obtaining of the occurrent matters of fact that they redescribe, we do not need to supplement our theoretical resources to account for laws of nature.

Conceptual motivation: Humeans can account for our grasp of the concept of a law. We can unproblematically grasp what it is for something to be an occurrent matter of fact, and we can unproblematically grasp what it is for some occurrent matters of fact to be configured in some way, so we can unproblematically grasp what it is for some law of nature to obtain.

Set against these motivations are objections from epiphenomenality, explanation and inference:

Epiphenomenality objection: Humean laws describe rather than governing. The laws themselves play no role in explaining the occurrent states of affairs that they efficiently describe.

Explanatory objection: Humean laws are incapable of explaining their instances. Explanation is asymmetric, and if the occurrent matters of fact being as they are explains the laws being as they are, then the laws being as they are cannot explain the occurrent matters of fact being as they are.

Inferential objection: Knowledge of the laws presupposes knowledge of the future, which gets things back to front. Laws are apparently used in science to project regularities from the past to the future, but we can only know that a Humean law is a law if we know enough about the future to know that the regularity it corresponds to will continue to obtain.

Naturally enough, friends of Humeanism about laws aim to defuse the objections to the view, while opponents of Humeanism aim to deflate the motivations for it; but underlying attitudes tend to remain polarized. In this chapter I am interested in developing a view that does justice both to the main motivations for Humeanism and to the main objections against it. Whether the resulting view should be classified as a version of Humeanism then becomes a moot point (albeit one to which I shall briefly return in section 4.8).

Consideration of the above motivations and objections allows us to formulate the following list of desiderata for a theory of laws:

Explanation: The laws must figure in explanations of occurrent matters of fact.
Knowability: The laws must at least sometimes be knowable using the sorts of evidence that we typically have.
Inference: There must be a clear rationale for using our best current assessment of what the laws are to draw inferences about the future.
Graspability: We must be able to explain how we acquire and deploy the concept of a law of nature.
Unspookiness: There must be no spooky metaphysical ingredients in the theory.

The entrenched dispute between Humeanism and anti-Humeanism might be taken to suggest that these desiderata are not jointly satisfiable; different parties to the debate typically accept as much and seek to give grounds for prioritizing some of the desiderata over the others. However, I think that the apparent impossibility of jointly satisfying the desiderata derives from an additional assumption that is often smuggled in unquestioned to the debate: that the dependence base for the laws is restricted to the actual world. Quantum modal realism rejects this assumption, and so it opens up conceptual space for a new style of theory of laws of nature.

In line with the previous literature on laws of nature, I shall formulate the various views of laws to be discussed in the form of universal generalizations concerning property instances. Of course, any realistic candidates for laws of physics will be substantially mathematized, but it is straightforward enough to replace 'all Fs are G' in what follows with mathematized functional dependencies between properties F and G.

We may start with the simplest possible Humean theory of laws, the naïve regularity view:

> *Naïve regularity theory:* It is a law that all actual Fs are G iff all actual Fs are G.

The naïve regularity view has well-known problems: in particular, it counts all true universal generalizations as laws. While we can tweak the view to get rid of vacuous universal generalizations by adding the conjunct 'something is F' to the right-hand side of the biconditional, we still face difficulties with non-vacuous universal generalizations which are true but are not laws. These include the notorious counterexamples of the spheres of gold and uranium. It is plausible that nowhere in the actual universe is there a gold sphere exactly one mile in diameter. It is also plausible that nowhere in the actual universe is there a sphere of uranium-235 exactly one mile in diameter. The latter absence, but not the former, we take to be a matter of law: if some technologically advanced species wished to build a giant sphere of gold, they might succeed, but since uranium-235 is fissile it is physically impossible (or close enough for the sake of the example) to construct one.

The most popular modification of the naïve regularity view is a Mill-Ramsey-Lewis (MRL) sophisticated regularity view (Mill 1843; Ramsey 1928; Lewis 1973, 1994), which imposes additional requirements on any universal generalization that is to count as a law. In particular, MRL theories add the requirement that the generalization be entailed by a best system, where a best system is (in Lewis's development of the idea) a set of theoretical statements that strikes an optimum balance between simplicity and strength. The idea is that the uranium sphere generalization will be entailed by such a system, but the gold sphere generalization will not be.

While the MRL approach to laws is popular, it faces a number of difficulties. It is very tricky to specify what counts as simplicity and what counts as strength, and there seem to be a number of possible ways to balance the two criteria off against one another. But, rather than delving deeper into the problems with the sophisticated regularity theory and into possible solutions to these problems, I shall explore a different approach to modifying the naïve regularity view.

We can reduce the number of universal generalizations which count as laws by *modalizing* the regularity theory: that is, by requiring lawlike generalizations to apply not merely to actual occurrent matters of fact but also to all possible occurrent matters of fact. In order to allow the account to properly handle non-fundamental laws of nature—to be discussed in more detail shortly—we also restrict the regularity account to cover only fundamental laws. After these two modifications, we obtain the following view:

Modalized regularity theory: It is a fundamental law of nature that all actual Fs are G iff all possible Fs are G.

The move from an over-simple theory to a modalized form of that theory may seem familiar: it arises in the debate over the metaphysics of properties on which we touched in section 1.6. As we saw there, Lewis adopted a modalized theory of properties to solve the problem of accidentally coextensive properties (Lewis 1986b), identifying properties with sets of their actual and possible instances. However, he did not make the analogous move in the theory of laws of nature. Why not?

One widespread concern about the modalized regularity theory is that it quantifies over merely possible individuals, but this cannot have been what dissuaded Lewis from adopting the theory. After all, Lewis's modal realism includes innumerable merely possible individuals, and he appealed to these individuals in the context of the modalized theory of properties. Instead, a plausible conjecture as to why Lewis modalized his theory of properties but not his theory of laws is that he recognized that a modalized regularity theory of laws would commit him to the non-contingency of the fundamental laws of nature, a consequence which he found intolerable.

The incompatibility of the modalized regularity theory of laws with the contingency of the fundamental laws is easy to see. The modalized regularity theory quantifies in an unrestricted way over possible individuals; only regularities that hold of all the possible individuals get to count as laws. But a contingent fundamental law would hold true of some, but not all, of the possible individuals. So the fundamental laws of nature must be non-contingent according to the modalized regularity theory.

Quantum modal realism dispenses with the assumption that fundamental laws are contingent, and thereby promises to rehabilitate the modalized regularity theory. In the context of quantum modal realism, metaphysically possible worlds are Everett worlds; hence the modalized regularity theory makes universal generalizations that are true of all things in all Everett worlds into non-contingent fundamental laws of each Everett world. Potential examples include the Pauli exclusion principle, as well as conservation laws for energy and momentum.

The presumption of contingency has a strong grip on contemporary theories of lawhood: Lewis made it an axiom of his original theory of laws (Lewis 1973: 73) without feeling the need to comment, and van Fraassen (1989) dismisses the view that the laws of nature are non-contingent as a 'minority opinion' unworthy of extensive discussion.[1] It is therefore unsurprising that making laws non-contingent has been seen as a significant problem for the modalized regularity theory, and—in the light of its other attractive features—it is unsurprising that some attempts have been made to reconcile the modalized regularity theory with contingent laws. In section 4.3, I describe and criticize one such reconciliation attempt, due to Robert Pargetter.

4.3 Pargetter's Theory

In what seems to me an underappreciated contribution, Robert Pargetter proposed a modal realist theory of laws of nature based on the modalized regularity theory (Pargetter 1984). Pargetter was impressed by the analogy between a modalized theory of properties and the modalized regularity theory sketched in section 4.2. This analogy motivated him to argue that modal realists should extend the modalizing move from their theory of properties to their theory of laws. However, in order to recover the presumed contingency of the fundamental laws, Pargetter added a new element to his version of the modal realist picture: primitive external equivalence relations between worlds, corresponding to the relation of *sameness of law*. While this addition recovers the contingency of fundamental laws as intended, I shall argue that it undermines the explanatory power of the resulting theory of laws of nature.

According to Pargetter's theory, laws are true universal generalizations over actual and possible individuals. However, the generalizations concerned are not completely unrestricted, as they are in the modalized regularity theory

[1] Although the dispositional essentialist tradition (e.g. Ellis 2001, Bird 2007) involves ascribing the laws a kind of necessity, the laws remain contingent in the sense that different properties governed (necessarily) by different laws could have been instantiated. The kind of necessitarianism that is incorporated into quantum modal realism and is at issue in this chapter is modal necessitarianism—a genuinely minority opinion! See section 4.3 for more discussion, and Bird (2004) and A. Wilson (2013b) for defence.

described in section 4.2 rather, the generalizations identified as laws of nature are generalizations over the actual and possible individuals in worlds that bear the distinctive new external equivalence relation to one another. This new external relation may be understood as playing the role of sameness of law in an implicit definition of the concept of law of nature. Like any other equivalence relation over worlds, the sameness-of-law relation partitions the worlds into distinct networks, such that every world in each network bears the sameness-of-law relation to every other world in that network but to no world in any other network,. It therefore permits the following theory of laws:

Pargetter's modalized regularity theory: It is a law that all actual Fs are G iff all Fs in worlds standing in the same-law relation to the actual world are G.

Pargetter posited the sameness-of-law relation as a new theoretical primitive. Like spatiotemporal relations in the Lewisian picture, the sameness-of-law relation is supposed to be a basic external relation that is not reducible to the intrinsic properties of the relata. The sameness-of-law relation is a novel theoretical commitment, but Pargetter argued that we can use it to account for natural properties so that his account has the same total number of theoretical primitives as Lewis's. Unfortunately, Pargetter's theory struggles with most of our desiderata; the problems are traceable to the status of the primitive sameness-of-law relation that he employs.

Take Explanation first. Absent some positive reason to think that worlds bearing the sameness-of-law relation to our own world form an especially salient contrast class, it is unclear how laws enable us to explain occurrent matters of fact. To put the point slightly differently: if some worlds contain Fs that are not G, what licenses us to ignore those Fs when we appeal to the laws to explain why some actual F is G? If we are wondering whether some actual F is G, then it is of course helpful to be told that all Fs in a set of worlds containing the actual world are G. But this is so regardless of whether the worlds in that set bear the sameness-of law relation to one another. The sameness-of-law relation does no additional explanatory work.

Knowability next. Without a reliable mechanism for acquiring knowledge of the sameness-of-law relations, Pargetter's theory makes laws unknowable. Since the sameness-of-law relation is an external relation, no amount of knowledge of the intrinsic properties of the relata will suffice for knowing whether it obtains. So, no amount of knowledge of the intrinsic properties of the actual world and of individuals within it will suffice for knowledge of the sameness-of-law relations in which the actual world stands to the other worlds. It is implausible to think that we can directly perceive or intuit the holding of the sameness-of-law relation. So Pargetter's theory seems to render it impossible in principle to obtain knowledge of the laws of nature.

The problem with Graspability is closely related to the problem with Knowability. Unlike the paradigmatic candidate case of external relations, spatiotemporal relations, we have no direct or even indirect experience of the sameness-of-law relation. Without such access, and without any ability to characterize it in terms of its consequences for the intrinsic properties of its relata (because it has no such consequences) we are left without any idea of how we could possibly have acquired the concept of sameness-of-law, and accordingly without any idea of how we could possibly have acquired any concept of law of nature that is characterized via the sameness-of-law relation. So Pargetter's theory makes lawhood ineffable.

Ungraspable relations are mysterious, and laws deriving from Pargetter's sameness-of-law relation consequently fail to satisfy the Unspookiness desideratum. The sameness-of-law relation that Pargetter posits is *sui generis*, and it is distinctively different from more familiar external relations (i.e. spatiotemporal relations) in that it holds between worlds rather than between objects within some single world. It is close to a paradigm case of a spooky metaphysical posit.

Only Inference is straightforwardly satisfied by Pargetter's theory. If we know that the actual world is a member of a set of worlds all of which have a certain property, then we are in a position to know that the actual world has that property. This success is cold comfort, though, in light of the theory's failure to handle our other desiderata.

Pargetter's view looks like a dead end. But introducing a primitive sameness-of-law relation is by no means the only way to respond to the contingency worry for the modalized regularity theory. We may instead reject the intuitions that motivate the contingency worry, and potentially also seek to explain them away in a manner compatible with the spirit of the modalized regularity theory. That is the strategy I recommend to quantum modal realists, and I expand upon it in section 4.4.

4.4 Laws in Quantum Modal Realism

The problems for Pargetter's modalized regularity theory can be traced to the primitive sameness-of-law relation that was introduced to restore the contingency of the fundamental laws. In the context of quantum modal realism the fundamental laws are non-contingent, so this motivation lapses. Quantum modal realists can therefore adopt Pargetter's core idea without complicating it in the problematic ways described in section 4.3. In this section, I will set out a modalized regularity view of laws that takes fundamental laws (as well as Fundamental laws) to be non-contingent. In section 4.5, I will argue that there is no deep theoretical cost to accepting non-contingency in fundamental laws, by first explaining how quantum modal realism still makes room for contingency of laws at the non-fundamental level and then defusing some intuitive motivations for contingency of all laws whatsoever.

Recall my formulation from section 4.2 of the modalized regularity theory of laws:

Modalized regularity theory: It is a fundamental law of nature that all actual Fs are G iff all possible Fs are G.

The kind of necessitarianism about laws that is implied by the modalized regularity theory is *modal necessitarianism*: the view that the actual laws are the laws of all metaphysically possible worlds. This is the strongest form of necessitarianism that has been taken seriously in the recent literature on laws of nature. Schaffer (2005) distinguishes modal necessitarianism from *nomic necessitarianism* and *causal necessitarianism*; both of the latter views allow for other metaphysically possible worlds with different laws of nature from our own world only insofar as those worlds instantiate different natural properties from our own. Modal necessitarianism, in contrast, maintains that the fundamental laws are the same in all metaphysically possible worlds and hence (on the assumption that a fundamental property exists at a world if and only if some fundamental law of that world specifies its behaviour) that the fundamental properties are the same in all metaphysically possible worlds.

The primary motivation for modal necessitarianism that I shall consider is a theoretical inference to the best explanation of the sort set out in section 0.2 of the introduction: the assumption of modal necessitarianism permits a better overall theory of laws than is possible without the assumption. However, modal necessitarianism can also be independently motivated; it allows for appealing accounts of our interest in natural necessity and of our practice of evaluating counterfactuals. I have argued this case elsewhere (A. Wilson 2013b), and I revisit some of those arguments in section 4.10 below.

Without quantification over possibilia, the modalized regularity theory collapses into the naïve regularity theory. Without modal necessitarianism, the modalized regularity theory delivers no substantive laws of nature. But with both quantification over possibilia and modal necessitarianism in the picture, the modalized regularity theory avoids the counterexamples to the simple regularity theory and it delivers a non-trivial set of fundamental laws of nature. The modalized regularity theory accordingly finds a very natural home in the context of quantum modal realism, which incorporates quantification over possibilia and seems inevitably committed to modal necessitarianism about the fundamental (and Fundamental) laws. This gives the modalized regularity theory strong naturalistic credentials; its primary ontological posits are drawn from within physics, rather than being special-purpose posits like Armstrong's necessitation relations between universals (Armstrong 1978, 1983).

It might appear that the modalized regularity theory is only able to deliver deterministic laws. But in quantum modal realism, the fundamental laws of Everett worlds are indeterministic. This is in fact fully compatible with the modalized regularity theory: we just need to build ascriptions of chances into the scope of our laws. For example, one candidate indeterministic law might be 'all electrons prepared in a z-spin eigenstate have a 50 per cent chance of being measured spin-up.' To simplify the discussion, I will not delve into the details of probabilistic laws and probabilistic explanation; for a systematically developed example of how probabilistic explanation may be accommodated in models of causal explanation like the one I have in mind, I refer the reader to Strevens (2008).

I imagine that anyone already convinced by quantum modal realism will find the modal regularity theory immediately congenial. For everyone else, still undecided about the merits of quantum modal realism, more work remains to be done to establish that the theoretical benefits of the modalized regularity theory (and of the broader quantum modal realist package) are worth the intuitive cost associated with accepting non-contingency of fundamental laws. So we must look in closer detail at those theoretical benefits. In the next three sections, I will apply the modalized regularity view in the context of quantum modal realism to formulate natural accounts of special-science laws and of *local laws*—laws which obtain in specific subsystems of the universe. Special-science laws can tolerate exceptions, and local laws can be contingent even when the fundamental laws of nature are non-contingent.

One variety of local laws in particular—contingent laws that build in specific values of the so-called 'constants of nature' which are the source of the notorious fine-tuning argument—will play an important role in the dialectic of this chapter, and in the overall quantum modal realist system. While the relevant physics is still unsettled, a number of theories being actively pursued by contemporary quantum cosmologists and quantum gravity theorists describe mechanisms that could potentially give rise to very different physical phenomena in different regions of the multiverse, or even within different regions of a single Everett world. I shall argue that these prospects largely undercut intuitive motivations for contingent fundamental laws. In light of the wide variety of physically possible phenomena that fundamental physics and cosmology seem set to deliver, and bearing in mind the powerful recombinatorial principles discussed in section 1.8, quantum modal realism manages to reconcile the non-contingency of fundamental laws with the substantial reliability of our modal reasoning.

4.5 Special-Science Laws

A major challenge for contemporary theories of laws of nature is the need to account for special-science laws. One of the most prominent criticisms of Lewis's best-system approach is that, focused as it is on the fundamental laws and their deductive consequences, it gives us no good account of laws within the special sciences. In response to these criticisms, authors including Schrenk (2006) and Callender & Cohen (2009) have offered alternative versions of a best-system approach which aim to do justice to special-science laws.

The modalized regularity theory avoids at least some of the problems with the Lewisian best-system theory of laws, since it does not depend in the same way upon a best system assessed for simplicity with respect to a fundamental language. But some problems remain: the modalized regularity theory gives only a criterion for fundamental lawhood, and it needs extending in order to accommodate non-fundamental lawhood. In addition, special-science laws seem to tolerate exceptions; to allow for that, we shall need to build some further flexibility into the account.

Fortunately, the modalized regularity theory is highly flexible: we can liberalize the theory along a number of dimensions. A first approach, motivated by the exception-tolerating nature of special-science laws, involves weakening the quantifiers that appear in the modalized regularity analysis of lawhood. Instead of using the ordinary universal quantifier, applied symmetrically on both sides of the analysis, we can make use of some weaker quantifiers applied to the right-hand-side of the biconditional. For example, consider the following weakenings of the modalized regularity theory:

Typicality Modalized Regularity Theory: It is a law that all actual Fs are Gs iff *typical* possible Fs are Gs.
Predominance Modalized Regularity Theory: It is a law that all actual Fs are Gs iff *most* possible Fs are Gs.

These weakened analyses have the consequence that some generalizations can be laws while not being true of the actual world. Such weakenings are a straightforward way to implement an exception-tolerating nature for special-science laws. But they fail to give us all that we might want from a theory of special-science laws; in particular, they do not help to characterize the laws of a particular special science. For that, we need to draw on some additional resources.

In order to characterize what I will call *domain-specific laws*—laws of particular special sciences, such as laws of chemistry or laws of biology—we can repeat the trick of varying the quantifiers used in the modalized regularity theory biconditional. However, rather than using some general but weaker quantifier, we can retain the universal quantifier and combine it with domain-specific qualitative quantifier domain restrictions. Recall that the left-hand-side of the modalized regularity analysis of lawhood is explicitly restricted to actual individuals, while the right-hand-side of the analysis is intended to be read as unrestricted. To account for special-science laws, we can impose additional qualitative domain restrictions on quantifiers on both the RHS and the LHS: where *Q* is a qualitative domain restrictor, it is a special-science law that all actual *Q* Fs are Gs iff all *Q* Fs are Gs.

It is easiest to see how this proposal works by looking at some examples. For example, consider the restrictor *with characteristic actions large relative to Planck's constant.* Systems with characteristic actions large relative to Planck's constant do not exhibit quantum-mechanical behaviour, so plugging this restriction into both sides of the modalized regularity theory biconditional gives us laws of the classical (non-quantum) domain.[2]

[2] It might be objected that it is not in fact impossible for such systems to display quantum effects, but merely vanishingly unlikely. In light of the decoherence-based solution to the measurement problem, and granting an appropriate role for context, this may turn out to be impossible after all; see section 4.8. But in any case, if need be we can appeal to weaker quantifiers as a fallback here.

Likewise, plugging in the restrictor *in regions of negligible spacetime curvature* gives us laws of special relativity, and plugging in the restrictor *with relative velocities small relative to the speed of light and in regions of negligible spacetime curvature* gives us laws of the non-relativistic domain. Given suitable qualitative restrictions corresponding to living systems, chemical systems, and mental states, plugging in the restrictor *living* gives us laws of biology, plugging in *chemical* gives us laws of chemistry, and plugging in *mental state* gives us laws of psychology.

Not every qualitative predicate, of course, combines with the modalized regularity analysis of lawhood to characterize an interesting special-science law. Disjunctive or gerrymandered predicates, for example, do not seem apt to generate genuine special-science laws. While it is certainly the case that all things that are both green and hungry are hungry, and it may be the case that all things that are either cheese or chalk are solid, neither of these generalizations is fit to count as a law of any respectable special science. Accordingly, an adequate modalized regularity analysis of special-science lawhood will need to impose further constraints on which qualitative predicates *Q* give rise to genuine laws. The obvious choice is to use a criterion of naturalness; the qualitative predicates involved must be *natural properties* or *natural kinds* in something like the sense of Quine (1969a) or Lewis (1983b). Discussion of naturalness in the context of quantum modal realism is taken up in section 4.9 below.

4.6 Local Laws

As well as qualitative restrictions on quantification over modal space, we can impose non-qualitative *de re* restrictions: for example, we can quantify over all Scandinavians or all *homo sapiens*. Without *de re* restrictions, we obtain general laws that apply to the whole of reality or to certain qualitatively characterized parts of reality. But with *de re* restrictions, we can characterize laws that apply to specific individuals or regions within reality. Such restrictions will turn out to provide a way of mitigating the most counterintuitive consequence of the modalized regularity view. Although quantum modal realism renders the fundamental laws of nature non-contingent, it makes room for an account of local laws which can effectively simulate contingent fundamental laws. This defuses the objection based on the counter-intuitive nature of the non-contingency of fundamental laws.

One particularly simple and obvious *de re* restriction is a spatiotemporal restriction: we restrict the quantifiers in both the RHS and the LHS to quantify over individuals in some spatiotemporal region and that region's counterparts in other worlds. This gives us 'local laws'. Suppose, for example, that we impose the *de re* restriction expressed by the phrase 'on Earth'. Then we have 'planetary laws' specific to Earth, determining such things as gravitational fields, magnetic fields, and horizon curvatures. For example (making all restrictions explicit) the modalized regularity theory gives us that it is a law that all actual Fs *on Earth* are Gs iff all possible Fs *on some counterpart of Earth* are Gs. According to this criterion, it is a law of Earth that unsupported objects in a vacuum accelerate downwards at $9.81ms^{-2}$.

There are as many candidate local laws as there are pairs of individual systems and counterpart relations. Of course, not all of these local laws will be interesting or useful; in particular, it is useful to impose a condition of naturalness on the restrictors that generate local laws. The obvious approach is to limit ourselves to local laws generated via a restriction to systems that are members of some genuine natural kind. For example, we may well have use for laws of psychology of an individual person X, generated by the restrictor 'psychological states of X'. We are unlikely to have any use for laws of psychology of a collection of disparate individuals, as generated by a restrictor of the form 'psychological states belonging to either X, Y, or Z', unless X, Y, and Z have something theoretically interesting in common—that is, unless this grouping corresponds to a natural kind or is metaphysically privileged in some similar way.

How do local laws and special-science laws interact? In principle, the restrictions which generate them are cross-cutting. We can apply the qualitative restrictions and the *de re* restrictions independently to generate special-science laws and local laws respectively, or we can combine them to generate local special-science laws. For example: laws of psychology of some individual, as generated by a qualitative restriction to psychological states and by a *de re* restriction to states of a particular individual, would be local special-science laws. However, this view may not be straightforward to apply, since apparently qualitative restrictions may themselves conceal hidden *de re* restriction. For example, a restriction to biological systems may turn out to be a *de re* restriction if 'biological' is not a purely qualitative predicate. This possibility is considered in section 4.7.

Exploring all the varied possible applications of local laws is a project well beyond the scope of this book. However, one particular application of local laws within quantum modal realism will play a major role in the overall dialectic; I turn now to consider it.

4.7 Parameterized Laws

The particular local laws so far considered were generated by restricting to systems that are relatively small and localized. However, any system to which we can refer can in principle be used as a restriction to generate local laws, including systems at the cosmological scale. A class of local laws which are especially interesting from the perspective of quantum modal realism are what I will call *parameterized laws*. These laws vary from region to region within the Everett multiverse, and accordingly they are contingent even in the presence of non-contingent fundamental and Fundamental laws.

There are two main types of mechanisms in contemporary physics which could give rise to parameterized laws. These are the string landscape model according to which parameters characterizing the whole spacetime of an Everett world are determined by a quantum-mechanically chancy process, and the eternal inflation model of quantum cosmology according to which a single Everett world may contain numerous 'bubbles' each of which contains a differently parameterized and effectively causally isolated spacetime. Although these are not the only models that could give rise to parameterized laws, I will focus on them; an alternative, and much more esoteric, potential mechanism would be the 'fecund universes' cosmological natural selection hypothesis of Smolin (1997).

Take first the case of string landscape cosmology, as proposed by e.g. Susskind (2005), following Bousso and Polchinski (2000). String landscape models provide one possible form that parameter-fixing dynamical processes could take. In the string landscape scenario, a unitary quantum process leads to an extremely complex superposition of the different possible compactifications of a Calabi-Yau manifold, with probabilities attached to each compactification; hence, in an Everettian implementation of this quantum chance process there would be Everett worlds corresponding to all possible compactifications. A widely discussed calculation (Douglas 2003) gives the number of possible compactifications at around 10^{500}; each possible compactification corresponds to a different combination of physical parameters, and each combination permits different physical phenomena within the resulting compactified spacetime. Each of the 10^{500} compactifications therefore corresponds to an enormous class of Everett worlds, with all of the worlds within each class sharing the same parameter values.

In the context of a string landscape Everettian multiverse, quantum modal realists may characterize parameterized *landscape laws* which impose a *de re* quantifier restriction to worlds which have the same values of the string landscape parameters as the actual world. The parameter values will then be fixed by the landscape laws of each Everett world, although they are left open by the fundamental laws. Landscape laws are contingent, and they encode the characteristic kinds of physical behaviour that are permitted within a given minimum of the string landscape potential.

Take next the inflationary approach to cosmology. Contemporary inflationary cosmology generically gives rise to *eternal inflation*, or *chaotic inflation*, where the rapid expansion of spacetime gives rise to effectively isolated regions of spacetime—*bubbles*—that have a Friedman-Lemaitre-Robertson-Walker (FLRW) metric. This metric ensures that, from within a bubble, spacetime appears homogenous, isotropic, and nearly flat. However, a single universe may contain myriad bubbles. There are several models of cosmological inflation which give rise to an enormous number of (even infinitely many) bubbles in each universe, but the details will be relatively unimportant for present purposes. What is important for the project of this chapter is that inflation be a quantum-mechanical process that has some probability of giving rise to bubbles with various different combinations of parameters characterizing physical phenomena playing out within each bubble. In the context of EQM, if there is some probability of a quantum-mechanical process giving rise to an outcome then there are Everett worlds in which that outcome occurs. So, given EQM and a theory of inflation which specifies a probability distribution over a range of combinations of values of cosmological parameters within each bubble, for every physically possible combination of parameter values there will be some inflationary bubble with that combination.

Since the property of being a bubble is an excellent candidate for a natural property in the context of inflationary cosmology, there will be local laws of nature generated by the restriction to systems located in some bubble or in its counterparts. When we take the modalized regularity theory and apply a *de re* restriction to counterparts of the bubble in which we ourselves are located, we obtain *bubble laws* characterizing behaviour which is physically possible within our own bubble. These bubble laws will vary from bubble to bubble, and the parameters that are observed locally will appear in the laws of our particular bubble. Bubble laws too are contingent, even in the presence of non-contingent fundamental and Fundamental laws.

Landscape laws and bubble laws are not mutually exclusive alternatives; there could be parameterized laws of both types that govern events in the actual visible universe. Whether there are laws of either type is an open question within contemporary theoretical physics and cosmology. If either type of law is vindicated by the progress of physics, or if some similar type of parameterized law is vindicated instead, quantum modal realists may appeal to these laws to capture contingent laws within their overall system. The contingency of parameterized laws goes a long way towards undermining the objection to the modalized regularity theory based on the non-contingency of the fundamental laws of nature. The contingency of the parameterized laws gives rise to regions of the multiverse, or bubbles within our own Everett world, which include physical phenomena that are very different from those with which we are acquainted. Indeed, there is every reason to think that these varied phenomena can do justice to the intuitions which motivate our belief in fundamental non-contingency.

Recent work on anthropic constraints on the value of cosmological parameters has begun to characterize the ways in which physical phenomena might vary in different minima of the string landscape, or in different bubbles. (For a sample, see Weinberg 1989, Tegmark & Rees 1997, Rees 2000, Barnes 2012). Defenders of the modalized regularity theory can point to this enormous contingency in the nature of physical phenomena to mitigate worries about fundamental non-contingency; this body of theoretical work in physics shows how the intuition that many different kinds of physical phenomena are possible is compatible with the thesis that there is no contingency in the fundamental laws. The role of anthropic constraints within quantum modal realism will be revisited in chapter 6.

The contingency of the parameters featuring in the landscape laws or bubble laws will give rise to contingency in the special-science laws which apply in particular minima of the landscape, or in particular bubbles. This is because changes in the parameter values ramify upwards through the ways in which phenomena at one level (say, the biological level) depend on phenomena at a lower level (say, the level of organic chemistry). So, a difference in the parameters that looks relatively slight from the perspective of fundamental physics may ramify up into an enormous difference from the perspective of biology or sociology. We have hardly begun to explore the consequences of different parameter values, so we know virtually nothing about the different chemistries or biologies that might be playing out even relatively nearby in the inflationary multiverse or in the string landscape. Nonetheless, we know enough to have a strong suspicion that there is more in the Everettian multiverse than is dreamed of—potentially, more even than can physically possibly be dreamed of—using the familiar macroscopic concepts that are adapted to our familiar macroscopic world.

I think the contingency of values of cosmological parameters, and the contingency of landscape laws and bubble laws which incorporate them, goes a long way towards undercutting the objection to the modalized regularity theory that trades on the intuition that the laws of nature could have been different. For all laws except the Fundamental and fundamental laws, contingency of laws is in factvindicated in the context of quantum modal realism. Exactly how far this contingency of parameterized laws extends is a question for scientific investigation, and answering it will depend on careful analysis of inter-theoretic relations as well as further progress in cosmology. Chapter 6 revisits these questions.

In section 4.8, I sketch a proposal for a contextual semantics for law-ascriptions that can pivot on the theories of special-science laws and of local laws just presented.

4.8 Context and Contingency

According to the modalized regularity theory of laws, to be a fundamental law is to be an exceptionless generalization over all actual and possible objects; to be a special-science law is to be a generalization over all actual and possible objects meeting some qualitative criterion; and to be a local law is to be a generalization over all actual and possible objects bearing some relation to a particular system or to its counterparts. In the light of this apparent proliferation of different concepts of law, what is it to be a law of nature *simpliciter*?

I propose that we think of being a law of nature as having a feature which these various different types of law have in common: *being a generalization over some actual and possible objects meeting some relatively natural criterion*. We can then account for our judgements about lawhood in particular situations by appeal to a contextualist semantics for ascriptions of lawhood. Lawhood, on this view, is a context-dependent business; when we ask after the laws of nature in some contexts, we may be asking after the fundamental laws or the Fundamental laws, while when we ask after them in other contexts, we may be asking after the laws of some special science or of some localized subsystem.

Simply declaring that something varies with context is philosophically unsatisfying. What we would like is a plausible account of how recognized mechanisms of contextual variation give rise to contextual variation in what laws are at issue. The modalized regularity theory allows for a straightforward account in terms of quantifier restriction, a very familiar application of contextualist semantics. 'There is no beer' can be true in my mouth so long as the fridge is empty, even if there is plenty of beer in my local pub. Context supplies an implicit 'in the house' or 'that is easily available' restriction on the quantifier domain. I suggest that a similar phenomenon may be in play with respect to domain-specific and local laws. Utterances of 'it is a law that all Fs are G' are to be assessed with respect to a domain restriction supplied by context. 'It is a law that objects accelerate downwards at 9.81ms^{-2}, is then typically true when uttered on Earth but false when uttered by someone on the moon. And, in numerous contexts where variation in cosmological parameters is not salient, we can truly assert that some contingent claims are laws: as I argued in section 4.7, this goes a long way towards undercutting the objection to quantum modal realism from the supposed non-contingency of laws of nature.

4.9 Naturalness

The necessity of fundamental laws in quantum modal realism challenges traditional ways of drawing the distinction between laws of nature on the one hand and the truths of mathematics or logic on the other. Lewis characterized laws as contingent generalizations, to be contrasted with the non-contingent generalizations of pure mathematics and logic, and he thereby avoided our difficulty. But if both the truths of pure mathematics and the Fundamental and fundamental laws of nature are non-contingent in quantum modal realism, how else can the distinction between them be drawn? Presumably, we do not want to conflate pure mathematics and fundamental physics completely, although this route might appeal to ontic structural realists such as Ladyman (1998) and French (2014).

The anti-Humean notion of 'governing' might have helped us distinguish fundamental physics from pure mathematics, if governing corresponded to some determination relation stronger than mere entailment on which we have an independent grip. However, it is one of the advantages of the theory of laws outlined in this chapter that it has no need to appeal to the mysterious notion of governing. We would do better to look elsewhere for demarcation purposes. As a schematic alternative, I suggest we can distinguish necessary laws from necessary truths of logic and pure mathematics by reference to what those truths are *about*. The fundamental laws of nature are about fundamental natural properties; less fundamental laws are about less fundamental natural properties; but the truths of pure mathematics and logic are not about natural properties at all. (Or perhaps the truths of pure mathematics and logic are about every property; as I will argue below, it makes no difference.)

This approach to discriminating between pure mathematics and logic on the one hand and fundamental physics on the other clearly requires discriminating between properties with respect to naturalness. Here there is solid precedent within metaphysics; Lewis (1983b) and Armstrong (1978) have convinced most of us that we need something like universals or objective naturalness. A metaphysics that is fully egalitarian with respect to properties is ultimately unable to characterize reality as having any non-trivial structure beyond the cardinality of the set of individuals. So we require some notion in the vicinity of naturalness. Fundamental laws of nature can then be characterized as just those universal generalizations which are *wholly about* perfectly natural properties. This solves our immediate problem, and demarcates fundamental laws of nature from pure mathematics and logic.

Of course, not all laws of nature that have been characterized in this chapter are fundamental laws, either in the big-F sense or in the small-f sense. Local laws are universal generalizations over the possibilia that are restricted by some criterion, qualitative or non-qualitative. The restriction allows us to characterize relative naturalness amongst the various local laws. The more natural the restriction, the more fundamental the law. To make this proposal fly, though, we will obviously need to say more about relative naturalness as well as saying more about perfect naturalness.

Lewis suggests that we characterize relative naturalness as complexity of definition in terms of the perfectly natural; Sider (2011) suggests a notion of 'structurality' expressed by a primitive operator which can attach to any linguistic items, including predicates; Schaffer (2009), Fine (2012b), and Bennett (2017) posit chains of grounding connecting some properties to other properties, and they thereby account for natural properties as the ungrounded properties. Each of these approaches would in principle be a suitable way to characterize naturalness for present purposes.

Once we have a suitable ideology of naturalness on board, we may still have further work to do in accounting for the distinction between laws of nature and principles of logic and mathematics. We need to make sure that principles of logic and pure mathematics do not count as being wholly about the relevant natural properties, while laws of nature do. Perhaps surprisingly, this is not obvious: depending on the exact theory of aboutness we prefer, we may say that (being *topic-neutral*) the truths of mathematics and logic are about all properties. If so, then a fortiori they are about perfectly natural physical properties. This is a consequence of a straightforward modal theory of aboutness—for example, it is a consequence of Lewis's theory (1988a, 1988b).

According to the Lewisian theory of aboutness, subject matters are partitions on the space of possible worlds. It is a trivial matter to replace Lewis worlds with Everett worlds in the analysis. For a proposition P to be wholly about some subject matter S, according to the resulting theory, is for the partition on the space of Everett worlds corresponding to *whether P* to be included in the partition on the space of Everett worlds corresponding to *how things are with respect to S*. On the resulting account, it turns out that logical and mathematical truths like 1+1=2 are wholly about any purported natural property, like having unit positive charge; the partition corresponding to *whether 1+1=2* is the trivial, maximally coarse-grained partition in which all worlds are included in a single cell, and this is included in every subject matter. Faced with this result, we have (at least) two options.

The first option, and the most straightforward one, is to slightly tweak the analysis of fundamental lawhood: we simply pick out as the fundamental laws those generalizations which are about fundamental properties, and only about fundamental properties. While attractively simple, this approach has significant limitations. In particular, all logical truths about particular fundamental properties will count as fundamental laws: for example, it will be a fundamental law that everything is either a quantum state or not a quantum state. This problem is not necessarily fatal; perhaps we can bite the bullet and accept a certain degree of revision in what we count as a fundamental law. It turns out there are just more fundamental laws than we had previously realized—albeit most of them, being logical or mathematical truths, are of limited interest. Still, it would be desirable to avoid this result if we can.

Fortunately, there is a second option. This is to reject the modal theory of aboutness, and to appeal to a more fine-grained theory of content for statements of laws which individuates them in terms of the epistemic scenarios they rule in and out, and which accordingly ties laws to specific natural properties in a way not matched by the truths of logic and pure mathematics. Recall that in section 1.4 we discussed Lewis's pluralism about content in the context of Lewisian modal realism, and that I endorsed a similar response in the Everettian context. This response undermines any demand to choose between fine-grained and coarse-grained contents corresponding to fundamental laws of nature in the quantum modal realist picture. Fundamental laws in quantum modal realism can be assigned coarse-grained contents which do not differentiate them from logical and mathematical truths, but they can also be assigned fine-grained contents which do so distinguish them. Both the coarse-grained and fine-grained contents unquestionably exist, given the resources of quantum modal realism: as I characterized these contents in section 2.4, they are (respectively) sets of Everett worlds and set-theoretic constructions out of sets of Everett worlds and sets of individuals in Everett worlds.

The fine-grained propositions corresponding to fundamental laws of nature will generally be what matters when it comes to accounting for our epistemic practices with respect to those laws, just as fine-grained propositions are what matters when it comes to accounting for our epistemic practices within the non-contingent domain of mathematics. All triangles are trilaterals, but at a given stage in a mathematical proof it may have been established that an object is triangular without it having been established that it is trilateral. Likewise, even if it is the case that no possible material object accelerates past the speed of light if and only if 1+1=2, it can still be the case that some rational epistemic agent believes the latter without believing the former.

4.10 Reviewing the Desiderata

It is time to return to the desiderata for a theory of laws that were described in section 4.2. In this section, I will argue that the quantum modal realist implementation of the modalized regularity theory of laws meets all of the desiderata.

The quantum modal realist implementation of the modalized regularity theory of laws satisfies the Explanation desideratum: it makes sense of how laws enable us to explain occurrent facts. In general, I suggest that laws enter into our explanations of occurrent facts by showing us how these facts counterfactually depend on other occurrent facts. This explanation-enabling role of laws is in turn anchored in the laws' status as objectively necessary. Why does our Everett world conform to the fundamental laws of nature?

Because *every Everett world does*; violations of the fundamental laws are objectively impossible. If violations of the fundamental laws are objectively impossible, then the demand to explain why they do not occur at our world lapses. Likewise, why does our Everett multiverse conform to the Fundamental laws? Because *every Everett multiverse does*; multiverses violating the Fundamental laws are objectively impossible. If violations of the Fundamental laws are objectively impossible, then the demand to explain why our Everett multiverse does not violate them lapses. It does not matter whether we regard these responses as providing positive explanations of why reality conforms to the laws, or merely as undercutting the need for any explanation of the kind requested. The key point is that, given the quantum modal realist implementation of the modalized regularity theory of laws, the objective necessity of the laws underwrites our general practice of explaining occurrent facts by holding the laws fixed in our counterfactual reasoning.

It is outside the remit of this book to explore in detail how it is that laws of nature enter into the explanation of contingent events; to properly engage with this question would require a substantial detour through the philosophy of explanation. My own preferred account of explanation is a generalized counterfactual theory which draws on the causal models used by interventionists such as Hitchcock (2001) and Woodward (2003) but which broadens the scope of the interventions that may be considered also to accommodate grounding explanations which cannot be understood in terms of the consequences of physically possible interventions. For further details, see A. Wilson (2018a, 2018b). The important features of this account for present purposes are the following.

First, the relata of the explanation relation are taken to be values of variables, or (equivalently) answers to questions; these relata may be represented via specific cells within a given partition of possibilities, modelled in the manner discussed in section 4.9 above. Second, explanations may be represented by structural-equation models which specify a pattern of directed functional relationships between values of one variable and values of other variables. The role of laws of nature is to determine which functional dependencies between variables hold—or, as I prefer to say, to *mediate* dependencies.[3] So the Explanation desideratum is in fact satisfied through laws

[3] Dependencies mediated by metaphysical principles, such as the principles of impure set theory, are classed as grounding rather than as causation by the criteria employed by A. Wilson (2018a, 2018b). Variables in grounding explanations will need to be understood as partitions over a space of scenarios.

mediating explanations that obtain between explanans and explanandum, rather than the laws themselves playing the role of explanans. This approach to laws and explanation is in the same general spirit as that of Skow (2016) and Schaffer (2016).

It is worth reiterating the structure of dependence that quantum modal realists envisage in the laws of nature. Individual events in a particular world will be causally explained through the mediation either of local laws, of special-science laws,or of fundamental laws. These fundamental laws are grounded, according to the modalized regularity theory, in the contents of all the Everett worlds. But in turn the contents of all the Everett worlds—the whole plurality of them—is collectively causally explained through the mediation of the Fundamental laws. Consequently, the ultimate causal explanation of any event occurring is mediated by the Fundamental laws governing the whole Everett multiverse. Ultimately, the Fundamental laws themselves mediate the explanation of why all Everett worlds, and a fortiori the actual Everett world, exist and have the properties that they do. In practice, our law-involving explanations are typically contrastive explanations of why one contingent event occurred instead of another—and these explanations will be mediated by fundamental laws, by special-science laws, and by local laws all of which govern individual Everett worlds.

The quantum modal realist implementation of the modalized regularity theory satisfies the knowability desideratum. The view I have outlined subsumes the epistemology of laws into modal epistemology more generally. And as I have argued in section 1.8, quantum modal realism allows for a powerful new naturalistic modal epistemology that makes facts about modal space directly discoverable by empirical scientific work. So long as we can come to know about the multiverse we inhabit, we can come to know the laws as characterized by the quantum modal realist implementation of the modalized regularity theory.

The quantum modal realist implementation of the modalized regularity theory satisfies the inference desideratum. In the simplest case: if we know that (unrestrictedly speaking) all Fs are Gs, then we are in a position to infer that an arbitrary actual F is a G. A similar point holds given Pargetter's theory: if all Fs in all worlds in a set including the actual world are Gs, then we are in a position to infer that an arbitrary actual F is a G. The quantum modal realist implementation of the modalized regularity theory inherits this virtue.

The quantum modal realist implementation of the modalized regularity theory satisfies the graspability desideratum. To grasp the notion of a law, on the view I have set out, requires only that we grasp universal quantification and that we grasp the notion of the actual world: that gigantic physical object which we all inhabit.

The quantum modal realist implementation of the modalized regularity theory satisfies the unspookiness desideratum. When we talk about the laws, we are talking about how occurrent facts are distributed across the Everett multiverse. The only metaphysical ingredients involved are the same sort of occurrent matters of fact as occur in the actual world. To slightly adapt a slogan of Lewis's: 'all there is in fundamental reality is a vast mosaic of local matters of particular fact, just one little thing and then another.'

With the successful satisfaction of our desiderata, the main goals of this chapter have been achieved. It remains to ask how we should classify the quantum modal realist implementation of the modalized regularity theory. The theory respects our original definition of Humeanism: laws are fully grounded in occurrent matters of fact. But these occurrent matters of fact are spread over multiple worlds; this feature in fact provides the 'modal force' characteristic of anti-Humeanism, since laws characterize not only how things are but how they must be. Explanations run in both the 'Humean' and 'anti-Humean' directions: we can run explanations both from occurrent matters of fact across multiple worlds to facts about the laws of a given world, and from facts about the laws of a given world to occurrent matters of fact within that world. Should the quantum modal realist implementation of the modalized regularity theory be classed as a reconciliation of Humeanism with anti-Humeanism, or as a vindication of Humeanism? For present purposes, it makes no difference at all.

4.11 Summary

The well-known MRL best-systems approach is not the only way to rescue Humean theories of laws from the problem of accidental regularities, and in the context of quantum modal realism it is not the best way. Quantum modal realism provides the materials for a modalized regularity account of laws that closely resembles Pargetter's account, with the main difference being that the fundamental laws turn out to be non-contingent.

In quantum modal realism we have Fundamental laws governing the most basic reality, the quantum state, alongside both fundamental laws and non-fundamental laws that each govern individual Everett worlds. The fundamental laws are regularities that hold across the whole Everett multiverse, while the non-fundamental laws are obtained by restricting our attention to various specific parts, or aspects, of the Everett multiverse. This quantum modal realist implementation of the modalized regularity theory solves a number of problems for Humeanism, while doing justice to a number of the Humean motivations. It provides powerful resources for characterizing local and special-science laws using only the ideology of the simple regularity view combined with the machinery of quantifier domain restriction. It allows for some powerful schematic explanations of why laws matter to us, and of how we can come to know them. It can be combined with familiar contextual quantifier-restriction mechanisms to provide a flexible semantic theory of law-ascriptions. In the light of these myriad virtues, it is of little import whether we think of the resulting theory of laws of nature as a hybrid of Humeanism and anti-Humeanism, or as an especially flexible form of Humeanism.

Does the explanatory success of this modalized regularity theory of laws of nature provide support for quantum modal realism? I think that to some limited extent it does; however, my main aim in the chapter has been to argue that Everettians have a clear and plausible story to tell about laws. In particular, I hope to have shown there is no persuasive argument against quantum modal realism that starts from the assumption that the fundamental laws of nature are contingent. The consequences that quantum modal realism has for laws of nature are not weaknesses of the view; if anything, they are among its strengths.

5
Indeterminacy

5.1 Introduction

Decoherence-based EQM encompasses enormously many macroscopic Everett worlds, but these worlds are non-fundamental: the multiverse is emergent. Individual Everett worlds are represented in EQM by decoherent histories, which are derived from semi-arbitrary coarse-grainings of the physical properties of the cosmos, and the whole collection of Everett worlds is accordingly non-fundamental. This derivative status of the Everettian multiverse leads to two key respects in which Everett worlds are indeterminate.

First, in contemporary decoherence-based EQM the space of Everett worlds is indeterminate with respect to the *number* of worlds it includes. Different coarse-grainings may each give rise to decoherent history spaces satisfying the decoherence conditions, and nothing in the theory picks out one over the other as the uniquely correct space of Everett worlds.[1] Second, Everett worlds are indeterminate in *nature*; a world may for example fail to determine which of two slits an electron travels through, if the electron wavefunction does not decohere in the process. These indeterminacies are directly linked: the more worlds there are, the less indeterminate each of these worlds is, and a precise 'interpretation basis' as endorsed by Everett (1957a) and Deutsch (1985) eliminates both indeterminacies completely, although by fiat.

While understanding the indeterminacy of the Everettian multiverse is primarily a challenge for philosophers of physics defending EQM, and hence not a part of this book's primary agenda, questions about quantum indeterminacy take on a distinctive new aspect for quantum modal realists. Indeterminacy in modal metaphysics is unfamiliar and unsettling. How can there fail to be a fact about how many possible worlds there are? How can possible worlds be qualitatively indeterminate? In this chapter I attempt to make sense of these ideas. I argue that the indeterminacies of both world number

[1] Section 2.3 provides further details of decoherent history spaces.

The Nature of Contingency: Quantum Physics as Modal Realism. Alastair Wilson, Oxford University Press (2020).

DOI: 10.1093/oso/9780198846215.001.0001

and world nature become comprehensible and unobjectionable once the non-fundamental status of worlds in EQM is properly appreciated. The main conclusion is that both indeterminacies ought to be regarded as epistemic or semantic in origin; however, indeterminacy of world nature also provides an interesting testbed for recent work on models of metaphysical indeterminacy and it may usefully be understood as a novel example of emergent ontic indeterminacy.

5.2 Indeterminacy of World Number

One of the most puzzling features of the multiverse of decoherence-based EQM is that the number of Everett worlds in the multiverse is *indeterminate*. Different coarse-grainings of the decoherence basis deliver different verdicts concerning world number, and none of these is physically or metaphysically privileged. The more fine-grained the partition employed, the more worlds there are counted as being, but only up to a point: the partition cannot be fine-grained without limit, for eventually the decoherence conditions will cease to be satisfied. Since the decoherence conditions are approximate, there is no precise point at which they cease to be satisfied. World number thus ends up indeterminate, in a sense which clearly deserves further scrutiny. Here is how David Wallace puts the point:

> There is no sense in which [chaotic] phenomena lead to a naturally *discrete* branching process... while a branching structure can be discerned in such systems it has no natural 'grain'. To be sure, by choosing a certain discretisation of (configuration-)space and time, a discrete branching structure will emerge, but a finer or coarser choice would also give branching. And there is no 'finest' choice of branching structure: as we fine-grain our decoherent history space, we will eventually reach a point where interference between branches ceases to be negligible, but there is no precise point where this occurs. As such, the question 'how many branches are there?' does not, ultimately, make sense. Wallace (2010a: 67–8)

Everettians have a variety of ways of expressing the thought that the number of Everett worlds is indeterminate. Sometimes counts of worlds are said to be 'interest-relative' (Saunders 1998: 313) or 'arbitrary conventions' (Saunders 2010a: 12); they are 'not well-defined' (Greaves 2004) or subject to 'some indeterminacy' (Wallace 2010a: 68), or they 'presuppose a piece of structure that is not present in the theory' (Greaves 2007a). The question 'how many worlds?' is said to be 'a non-question' (Wallace 2010a: 68), with 'no good answer' (Saunders 2010a: 12) or simply with 'no answer' at all (Wallace 2010a: 68). World number 'has no categorical physical significance; it is not part of what is really there' (Saunders 2005: 235).

Some of these comments evoke a hard-line option of rejecting the question 'how many Everett worlds?' as literally nonsensical. This kind of rejection of an apparently meaningful question is reminiscent of the rejection by Ladyman and Ross (2007) of questions about the exact boundaries of macroscopic objects. In both cases, it is difficult to know exactly how to understand the consequences of the rejection. If mountains have no boundaries, but it is also not the case that everything overlaps with every mountain, it follows that there are no mountains. If there is no number n such that n is the number of branches, it follows that there are no branches. Of course, Ladyman and Ross do not deny that there are mountains; and Wallace, Greaves, and Saunders do not deny that there is branching. Our puzzle, then, is how to make sense of how it can be that there is branching despite there being no n such that n is the number of branches.

The key to resolving the puzzle, I think, is to grant that there is *something* defective about the question 'how many Everett worlds?' but to deny that this requires it to be meaningless, or nonsensical, or to lack a true answer. We can instead say that the number of worlds is *bivalently indeterminate*: that there is some n such that n is the number of worlds, but there is no n such that n is determinately the number of worlds. The question 'how many worlds?' is defective in the sense that it has no determinate answer, even though (determinately) it is meaningful and thus (determinately) it does have a correct answer. This sort of response is hinted at by Wallace's talk of 'indeterminacy' and by Greaves's talk of 'vagueness', but it has not to my knowledge anywhere been made explicit; it also appears to be in tension with the use of phrases like 'non-question'. I therefore propose it not as interpretation of Greaves, Wallace and Saunders but as a (friendly) amendment to their strategy, designed to make decoherence-based EQM more palatable to mainstream metaphysicians.

The bivalent indeterminacy treatment of questions about world number is only satisfying if we can give an appropriate account of determinacy. The contemporary literature on vagueness provides various frameworks for thinking about determinacy and indeterminacy that allow us to retain classical logic. I will discuss only the most prominent approaches, each of which makes use of the notion of *admissible precisifications*. A precisification is, as the name suggests, a way of making precise a vague expression. The characterization of admissibility varies from theory to theory, as do the nature of the precisifications. What the different theories have in common is a conception of determinacy as truth (or falsity) on all admissible precisifications, and indeterminacy as truth on some but not all admissible precisifications.

Epistemicist approaches to vagueness[2] say that some precisification is picked out by patterns in our community's global linguistic usage as objectively correct, but that it is typically unknowable (for distinctive reasons) which precisification is correct. According to epistemicism, admissibility is epistemic in nature, and indeterminacy is a kind of ineliminable ignorance.

Supervaluationist approaches[3] in contrast hold that admissibility is semantic in nature: admissible interpretations are those whose correctness is not ruled out by whatever meaning-determining facts or principles (such as explicit or implicit conventions) govern language use. For supervaluationists, indeterminacy derives from 'semantic indecision' (Lewis 1986b: 213).

Let's see how these two views apply to the indeterminacy of world number present in decoherence-only versions of EQM. We treat 'Everett world' as a vague expression, and consider various ways of making it precise. Each of these ways corresponds to some particular coarse-graining of the decoherence-based decomposition of the quantum state into components. The coarse-graining we choose is subject to constraints: if it is too fine-grained, then interference between the components will become non-trivial and decoherence will be lost. Conversely, if we make it too coarse-grained, there will cease to be any branching at all. In addition, no coarse-graining gives rise to non-integral values for world number. We can therefore place two extremely minimal constraints on admissible precisifications for 'the number of Everett worlds' in any episode of branching of a finite system: every admissible precisification must correspond to some natural number of Everett worlds, and this natural number must be greater than 1.

The constraints on precisifications described in the previous paragraph ensure the right results in the case of questions about world number. According to both epistemicism and supervaluationism, it makes perfect sense to say that there is some N such that N is the number of worlds, but it is

[2] Williamson (1996) gives the canonical contemporary defence of epistemicism.

[3] Fine (1975), Lewis (1982), and McGee & McLaughlin (1995) give influential (though differing) versions of supervaluationism. I will mostly be discussing McGee & McLaughlin's version, since (like epistemicism) it allows for the preservation of bivalence.

indeterminate which N is the number of worlds. Despite the indeterminacy of world number, it remains determinately true that there is branching.

The bivalent indeterminacy strategy thus allows us to capture a number of plausible thoughts about branching. We can say that determinately there is branching, and that branching is determinately happening massively and on very short timescales, and we can evaluate competing claims about whether branching is into finitely or infinitely many worlds. If some argument from determinate premises concerning the Everett multiverse goes through on every admissible precisification of the decoherence basis, then on the bivalent-indeterminacy model the argument's conclusion will hold determinately even though world number remains indeterminate. Problematic consequences of indeterminacy of world number are contained.

As it happens, I expect that most Everettians, if they favour my account of world number indeterminacy in terms of a precisification-based theory of vagueness, would prefer a supervaluationist reading of the indeterminacy involved. Epistemicism is widely considered implausible even in classical cases of vagueness—it is simply hard to believe in all those unknowable precise facts of the matter—and I see no particular reason why epistemicism should be any more or less plausible in the case of world number than it is in classical cases of vagueness. In any case, I will not attempt here to adjudicate between an epistemicist and a (bivalence-preserving) supervaluationist interpretation of the determinacy operator in the context of questions about world number; the debate turns on delicate questions in meta-semantics. What matters for present purposes is that both views allow us to preserve classical logic and semantics in full generality when describing the multiverse of EQM.

The claim that world number is vague could be resisted in a number of ways. For example, van Fraassen (1980) suggests as a constraint on the usefulness of vague predicates that there be clear cases and clear counter-cases. While there are clear counter-cases for 'is the number of worlds' (for example zero) there are no clear cases—no numbers *n* such that determinately *n* is the number of worlds. Van Fraassen's constraint seems to be overly restrictive, though. For example, consider the evidently vague predicate 'is the smallest large number'. We might think that zero is a clear counter-case: it is as far from large as anything could be. But there are no clear cases—no number is such that it is clear that it is the smallest large number. Van Fraassen's suggested characterization of vagueness accordingly looks inadequate. Vagueness should instead be thought of as the phenomenon of *borderline cases*; that characterization (which is favoured by Williamson 1996) allows world number to count as indeterminate.

Wallace prefers to say that it is not vague but *arbitrary* how many worlds there are (Wallace 2012: 101). However, this seems to have some implausible consequences. Arbitrariness is in the primary sense a property of choices, or of decisions: if a choice is arbitrary, then it is *up to us* how we make it. But it is not, in any coherent sense, up to us how many Everett worlds there are. (It may in some sense be up to us what we mean by 'Everett world': but given any determination of what is to count as a world, it is not up to us how many such objects there are.) Of course, it makes sense to say that for practical purposes we must pick some coarse-graining to analyse a particular physical interaction, and that the choice we actually make in any given situation is arbitrary; but this is perfectly compatible with (indeed, it can plausibly be explained by) the thesis that world number is indeterminate. Knowing that something is indeterminate but picking a single precisification to work with invariably involves making an arbitrary choice. Pressed on this point, I suspect Wallace would retreat to the claim that questions of world number—and consequently questions of person number and mind number, *inter alia*—are literally meaningless in the context of EQM. Although some will perhaps find a stable resting place here, this position conflicts too much for my liking with the Quinean methodology of section 0.4 of the introduction, which seeks reconciliation wherever possible between the scientific and manifest images, and which values minimal upheaval to classical logic and semantics. I will accordingly concentrate on less radical philosophical moves when explicating the metaphysical commitments of quantum modal realism.

The claim that world number is indeterminate is already unsettling enough. Quineans might well worry how we give adequate identity conditions to entities indeterminate in number; and prima facie such entities might give rise to a version of Evans's argument against vague identity (Evans 1978). There are many reasons of this general sort to doubt that vagueness is possible in fundamental reality. Everettians, however, have a powerful response. It is that, in EQM, worlds (along with the stars, people, and minds that they contain) are not fundamental; rather, Everett worlds and their contents are emergent phenomena. Saunders (1993, 1995) and Zurek (1993) have emphasized that there need be no vagueness at the more fundamental level, the level of the quantum state. Wallace (2010a) argues forcefully that approximation and indeterminacy are characteristic of emergence in the non-fundamental sciences. According to this line of thought, EQM is altogether unexceptional in using vague terms in its explanations.

The question of whether the number of worlds involved in some branching interaction is infinite or finite is closely connected to questions about the nature of the decoherence basis, and to issues in quantum gravity. In elementary presentations of EQM, it is usual to work with finite-dimensional Hilbert spaces; but realistic models of complex quantum systems frequently make use of continuous Hilbert spaces. It is an open theoretical question in cosmology whether the number of distinct possible histories of the universe is finite or not.[4] It is beyond the scope of this book to investigate whether branch number is typically finite or infinite in our best current theories in quantum cosmology: but however this question is resolved, the bivalent-indeterminacy approach is capable of modelling indeterminacy of branch number.

The picture which results from combining decoherence-based EQM with a precisificational theory of indeterminacy can be usefully compared with Deutsch's 1985 proposal, described in section 2.2. Instead of Deutsch's continuous infinity of diverging worlds, the picture I have outlined introduces an indeterminate number of worlds, leaving open the possibility that the number of worlds might be determinately finite. Where Deutsch introduced apparently unwarranted precision into the foundations of the theory, the indeterminacy approach accepts imprecision in the emergent multiverse and models it using the well-understood logical properties of the 'determinately' operator studied in precisificational theories of vagueness. It is determinate that there is some positive natural number, greater than one, of Everett worlds that diverge after any interaction; it is indeterminate what that number is.

Precisificational theories of vagueness typically allow for (but do not necessitate) context-dependency in the truth-value of vague statements, and in which precisifications are admissible. Should we admit context-dependence, in addition to indeterminacy, in statements about macroscopic reality in EQM? There is prima facie reason to think we should: just as it is implausible to think that a gradable adjective like 'hairy' expresses the same property in all contexts, it is implausible to think that the same degree-of-coarseness-of-grain is appropriate for determining world number in all descriptive contexts. We normally think that gradable adjectives may express different properties on different occasions of use, depending on contextually determined standards; we may if we wish apply a similar analysis to Everettian branching. Let us briefly explore how such an approach would work.

[4] For example, it is claimed that the Bekenstein bound requires the Hilbert space to be finite-dimensional for any system with finite spatiotemporal extension and finite energy. This (controversial) requirement is known as the 'holographic principle'. See, e.g., Bousso (2002).

Consider first the simplest possible case, of a spin measurement. Since there are two possible outcomes—Up and Down—in order to capture the required structure of the case we must use a coarse-graining which divides the state space into at least two worlds. A coarse-graining which washed out the difference between Up and Down altogether would be inadequate to our descriptive aspirations. Now imagine that the spin measurement is followed by a measurement of spin along an orthogonal axis. To adequately model the combination of the two measurements would require four worlds, one for each possible alternative sequence of results. So in the single-measurement case, there would determinately be at least 2 distinct worlds; in the dual-measurement case, there would determinately be at least 4 distinct worlds.

Moving from toy models to more realistic models, we will want to do justice to other degrees of freedom of a system. In a very detailed model of some measurement, we would need to include—for example—the interaction of the measurement apparatus with the cosmic background radiation. To do justice to these factors, we would need to use much finer coarse-grainings of the state space; these finer coarse-grainings would correspond to a vastly larger number of branches. We typically need thousands of worlds in our decoherent history space to do justice to interesting dynamical phenomena, even for very simple systems such as a particle in a bubble chamber. This suggests the following thought: perhaps the answer to the question 'how many Everett worlds branch from this interaction?' is context-sensitive,[5] with the parameter of contextual variation being *the salient level of descriptive detail.* According to this picture, in a context of a conversation where only the up-or-down result of a spin measurement is at issue, it might be true to say that the measurement results in two distinct worlds; in the context of an attempt to model some particular measurement instrument in maximum detail, it might be true to say that thousands of distinct worlds result from switching it on, with the instrument in subtly different states in each world.

This proposal does not place tight constraints on world number. Typically there will be a huge range of different coarse-grainings of the outcome space between the minimally adequate one which distinguishes all and only the structure we are interested in, and the level of fine-graining at which decoherence starts to break down. So, even allowing for contextual

[5] One might attempt to model the same phenomenon using interest-relativity rather than context-sensitivity. For the sort of reasons set out in Cappelen & Hawthorne (2009), I am doubtful about truth-relativization as an alternative to contextual variation.

variation, there will still be significant indeterminacy affecting questions of world number.

The contextualist proposal just described remains speculative, and I want to emphasize that quantum modal realists do not need to rely on it. The picture of branching that may be extracted from the writings of Everettians such as Saunders and Wallace, when understood as incorporating indeterminacy of world number, has it that in all contexts the number of worlds is determinately enormous. While we may model some interaction as resulting in two worlds, for Saunders and Wallace this is nothing but a modelling idealization. Even if the number of worlds is finite, in no context will it literally be correct to say that the number of worlds emerging from any but the simplest interaction is fewer than many thousands. Of course, the contextualist approach does not completely avoid this picture of extremely extensive branching; in some contexts, the most fine-grained space of worlds available will be the most salient. The only question I leave open here is whether it is ever literally true in any familiar context that the number of Everett worlds is small, or whether the coarse-grained partitions of Everett worlds that give rise to those lower world-counts are always mere idealizations.

Let me sum up the key points of this section. It remains an open cosmological question how many Everett worlds there are in the multiverse; but their number—if finite—is indeterminate. More generally, the number of worlds branching from any quantum interaction—if finite—is indeterminate. However, we can give straightforward accounts of this indeterminacy of world number, and we may rest assured that there will be enough determinacy in the space of Everett worlds to render them usable for the analytic project of quantum modal realism.

5.3 Indeterminacy of World Nature

It is not just world number that is indeterminate in decoherence-based EQM. Everett worlds are themselves indeterminate with respect to the microscopic physical quantities that they instantiate. This indeterminacy of world nature, as I shall call it, is tightly linked to the indeterminacy of world number. The more fine-grained our partition of a consistent history space, the more histories there are and the more determinate each history is—up to the point at which the decoherence condition is not satisfied. It is a vague matter where this point is located. However coarsely or finely we grain a decoherent history space, events within individual Everett worlds will exhibit some (indeterminate) degree of indeterminacy in their properties. It is determinate that there is qualitative indeterminacy in the worlds, but it is indeterminate exactly how much indeterminacy there is.

As with indeterminacy of world number, indeterminacy of world nature is compatible with determinacy with respect to many specific questions about world nature. Even if a particle lacks a precise position, it can still be determinately somewhere in the box. Even it is indeterminate whether the photon went through the upper slit or the lower slit, it can still be determinate that it went through one or the other. Since decoherence occurs much more rapidly with respect to systems with many degrees of freedom, determinacy will tend to be associated with macroscopic states of affairs, with extensive indeterminacy restricted to microscopic states of affairs. Where superpositions are magnified to large length scales—as happens in experiments probing the quantum-classical divide via superpositions of large molecules (e.g. Arndt et al. 1999)—Everett worlds are indeterminate with respect to properties of large systems.

Indeterminacy of world nature is unsettling, especially when it extends to larger length scales as in the numerous experimental displays of quantum interference. The two-slit inference effect was a strange and unexpected discovery, and no interpretation of quantum mechanics fully explains away this strangeness. Still, the Everettian implementation of indeterminacy of world nature minimizes the strangeness in one key respect: fundamental reality remains perfectly precise. As section 2.2 explained, contemporary Everettians assimilate the indeterminacy in Everett worlds to the more general indeterminacy of the entities and processes described by non-fundamental theories.

The properties of macroscopic systems that take on indeterminate values are emergent rather than fundamental properties. But if we understand EQM in the quantum modal realist way then the indeterminacy of world nature does inevitably entail indeterminacy in the nature of the actual world. There is no way of coarse-graining the history space so as to make the actual Everett world fully determinate, and so it is determinate that there is some indeterminacy in the correct description of the actual world. Quantum modal realism incorporates a precise modal reality, but an indeterminate actuality.

These distinctive features of quantum modal realism have the consequence that the usual contrast between representational and worldly indeterminacy ceases to apply straightforwardly. The indeterminacy of the actual world is representational, in the sense that it depends on a semi-arbitrary choice of coarse-graining; but it is also worldly, in the sense that a complete description of the actual world fails to eliminate this indeterminacy. Consequently, in the context of quantum modal realism, indeterminacy in world nature may be thought of as a naturalistic form of *metaphysical indeterminacy*, sometimes known as *ontic vagueness*.

Metaphysical indeterminacy has been modelled in various different ways over recent years, and it is interesting to see how this work might be applied to help understand the emergent metaphysical indeterminacy present in quantum modal realism. I will briefly discuss two models: the precisificational model of Barnes & Williams (2011), and the determinable-based model of J. Wilson (2013) and Wolff (2015). The aim is not to commit to either of these models, but to give a sense of how they play out in the context of quantum modal realism.

According to the Barnes–Williams proposal, precise metaphysically possible worlds correspond to distinct 'ontic precisifications' of how things are, with an ontically vague reality represented not by some single actuality but by a family of *multiple actualities*: worlds that do not determinately misrepresent how things are. A proposition is metaphysically determinately true if it is true with respect to all of the candidate actualities. The model is compatible with an older-style supervaluationist approach in which truth *simpliciter* is identified with truth at all candidate actualities. It is also compatible with the bivalent supervaluationist approach discussed above in which truth is truth at some preferred actuality, although it remains indeterminate which actuality is preferred. Both of these versions of the ontic precisification approach are live options for understanding indeterminacy of world nature in EQM.

The application of the ontic precisification model to the quantum setting has been criticized by Darby (2010) and by Skow (2010) (see also J. Wilson 2013 and Wolff 2015). Darby and Skow argue that representing quantum indeterminacy by multiplicity of precise histories does not do the phenomenon justice, since those precise histories are not themselves possible histories according to quantum mechanics. In the Everettian context, the point is the obvious one that since each Everett world is itself qualitatively indeterminate, indeterminacy of world nature cannot be wholly reduced to indeterminacy about which Everett world is actual. Any precise entities that could play the role of multiple actualities in an ontic precisification approach would not be Everett worlds. Can we nonetheless find suitable candidates for multiple actualities within EQM? One prospect is that they might be identified with quantum consistent histories as described in section 2.3. In order to play the role of ontic precisifications, the consistent history space in question would need to be maximally fine-grained. The decoherence conditions fail for these fine-grained consistent histories, so they are not dynamically decoupled from one another and quantum modal realists ought not to regard them as representing genuine alternative possibilities. Still, these consistent histories may be apt to play a different role in the metaphysics of quantum modal realism: the role of ontic precisifications in a Barnes–Williams-style model of metaphysical indeterminacy.

I will not delve further in this book into the details of how the ontic precisification model might be developed, but a number of important details remain to be resolved.[6] An alternative approach to the metaphysical indeterminacy of Everett worlds is appealingly simple by comparison to the precisificational model. Wilson and Wolff characterize quantum metaphysical indeterminacy without appeal to precisifications, as consisting in the possession of a determinable property without the possession of a determinate of that determinable. To illustrate, a particle in a box with an indeterminate position may possess the determinable property of *being located within the box* even if it does not possess either the property of *being located in the left half of the box* or the property of *being located in the right half of the box*. While in typical cases we take possession of a determinable property to necessitate (and perhaps to be grounded in) possession of a corresponding determinate, the determinable-based model of quantum metaphysical indeterminacy denies that this principle generalizes to the possession of arbitrarily precise microscopic properties by quantum systems.

There are several advantages to the determinable-based approach in the context of quantum modal realism. It avoids the need to appeal to a larger space of consistent histories in addition to the decoherent history space that is taken to represent the Everett worlds. Everett worlds are indeterminate, in the determinable-based approach, by being not fully specific, rather than by being associated with a collection of consistent histories which collectively represent them. Bivalence is preserved, without the need to introduce a 'preferred precisification' into the model; instead of a truth-value gap, we have only an absence of instantiation of any determinate property. The determinable-based model hence seems a more natural fit with the coarse-grained Everett worlds represented by a coarse-grained decoherent history

[6] A different kind of metaphysical supervaluationism involves appeals to partial situations instead of fully detailed worlds: see Darby & Pickup (forthcoming).

space. While the model seems to me promising, detailed development of the determinable-based approach to quantum indeterminacy must await another occasion.[7]

5.4 Summary

I have not aimed here to resolve every issue around indeterminacy in the Everett multiverse; understanding indeterminacy in the microscopic properties of quantum systems is a challenge for EQM in general, and one that is largely independent of quantum modal realism. My aim in this chapter has been to show, in outline, how quantum modal realism can accommodate whatever indeterminacy there may be in the Everettian ontology. I have argued that a bivalent precisificational account of the indeterminacy of world number enables us to identify various determinate features of the multiverse of diverging EQM and hence helps to underwrite the theoretical resources on which quantum modal realism relies. This approach alleviates worries about the coherence of EQM by connecting the framework used to explicate it to the well-understood machinery of precisificational indeterminacy. If the arguments of this section are correct, indeterminacy of world number need not be treated as different in kind from more familiar cases of vagueness. Likewise, potential worries about indeterminacy of world nature in EQM can be assuaged by adapting of one of the various models of metaphysical indeterminacy that has been proposed in the recent literature.

Much of the mystery around metaphysical indeterminacy is circumvented in quantum modal realism, since the indeterminacy does not extend to the fundamental ontology of EQM. Both indeterminacy of world number and indeterminacy of world nature are indeterminacies with respect to non-fundamental reality, and accordingly the indeterminacy of the emergent Everettian multiverse may be assimilated to the indeterminacy of emergent ontology more generally. Fundamental reality remains fully determinate in quantum modal realism: all that is indeterminate is the number and nature of the non-fundamental macroscopic Everett worlds, one of which we inhabit.

[7] Thanks to Robert Michels for helpful discussion of this material.

6
Anthropic Contingency

6.1 Introduction

Quantum modal realism is a theory of what really can happen. Yet not everything that can happen can be observed to happen. Some of the worlds in the Everett multiverse, of course, do contain observers observing reality to be some way; our own Everett world is one such world. Other worlds in the Everett multiverse contain no observers (whether because conditions in them are inhospitable to life, or because by chance life never occurs in them) but these worlds exist nonetheless. This disparity between Everett worlds that host observers and those that do not has epistemic consequences: it gives rise to an *observation selection effect*. In short, we ought not be surprised that we do not observe any course of events which cannot be observed. This simple point makes a profound difference to how we evaluate the evidential import of the fact that we observe conditions in the actual Everett world to be hospitable to life.

In this chapter my goal is to explore the scope and limits of *anthropic reasoning* in the context of quantum modal realism. Anthropic reasoning seeks to factor observation selection effects into the evaluation of the epistemic consequences of our observing the kind of world that we see around us. In particular, anthropic reasoning provides a powerful response to the well-known *fine-tuning argument* which uses the apparent extreme fragility of life with respect to variations in large-scale features of the cosmos to support the conclusion that the cosmos was intentionally designed to permit the evolution of life. I shall argue that quantum modal realism provides a new potential way to vindicate anthropic reasoning and undercut the fine-tuning argument. This particular vindication of anthropic reasoning is unavailable in the context of any approach to quantum theory other than EQM, and hence it highlights a surprising way in which the choice of interpretation of quantum mechanics bears evidentially on the question of whether physical reality was designed.

The viability of the overall quantum modal realist picture does not depend on any of the arguments in this chapter. Still, it is instructive to see how the

The Nature of Contingency: Quantum Physics as Modal Realism. Alastair Wilson, Oxford University Press (2020).

DOI: 10.1093/oso/9780198846215.001.0001

quantum modal realist reconceiving of the nature of contingency has broader ramifications not just for the foundations of metaphysics but for epistemology and the philosophy of religion. In section 6.2 I reprise the standard dialectic of the fine-tuning argument, and sketch the debate over whether multiverse hypotheses can underwrite an anthropic explanation of fine-tuning as an alternative to design. Section 6.3 explains a role for multiverse hypotheses that differs from that standardly considered: multiverse hypotheses that have some independent support from physics may serve as *undercutting defeaters* with respect to the fine-tuning reasoning. Section 6.4 explores different types of multiverse and assesses the extent to which they are capable of undercutting fine-tuning, and section 6.5 focuses specifically on a potential route to undercutting fine-tuning in the context of EQM which is unavailable in any one-world approach to quantum physics. Section 6.6 concludes that, surprisingly enough, interpretation of quantum physics is evidentially relevant to the question of whether physical reality was designed.

6.2 The Fine-tuning Argument

One of the most striking features of contemporary cosmology is that our best theories include a number of parameters that are fine-tuned with respect to life. This term has various uses; here I shall use it to mean any physical variable such that i) the value of the variable is not explained within our best theories and ii) moderate variations in that variable give rise to cosmological models where complex life is not physically possible. Such variables are sometimes called constants; in this chapter I will refer to them as parameters, since whether they are in fact constant is part of what is disputed.

A number of apparently fine-tuned parameters feature in modern cosmology. Martin Rees (2000) identifies six dimensionless parameters: the dimensionality of space ($D=3$), the ratio of the strengths of the gravitational and electromagnetic force ($N \approx 1036$), the nuclear efficiency of fusion from hydrogen to helium ($\epsilon \approx 0.007$), the density parameter characterizing the mass distribution in the universe ($\Omega \approx 1$), the cosmological constant ($\lambda \approx 10^{-122}$) and the ratio of the gravitational potential energy of a galaxy cluster to the mass-energy of that cluster ($Q \approx 10^{-5}$). For a detailed review of the associated physics, see Barnes (forthcoming). While some of these parameters appear to be more fine-tuned than others, and future theories may yet explain the values of some of them, when taken together the fact that the combination of values of all of these parameters appears so delicately poised to permit life is very striking, and it at least motivates the search for some underlying explanation.

According to the notorious fine-tuning argument, the evidence of cosmological fine-tuning provides confirmation for the proposition that there is a designer. The basic thought is that fine-tuning evidence would be much less surprising if there is a designer than this evidence would be if there is no designer. Fine-tuned universes fit the design hypothesis better than they fit the no-design hypothesis, and accordingly a discovery of fine-tuned parameters characterizing our universe tends to provide evidential support for design.

Anthropic reasoning has sometimes been deployed to block the fine-tuning argument, by enabling us to resist the thought that fine-tuning evidence would be surprising if there was no designer, and hence to deny that there is a disparity in surprisingness. But the anthropic response to the fine-tuning argument has been influentially criticized. In John Leslie's vivid analogy (Leslie 1989), you ought to be surprised to find yourself alive after a reliable firing squad has attempted to shoot you (and you may reasonably infer some unknown cause of their all missing) even if you wouldn't have been around to be unsurprised if they had successfully carried out their task. A reliable firing squad all missing is just (we may say, without yet committing to any particular analysis of this notion) intrinsically unlikely. The problem with the anthropic response is that the occurrence of fine-tuned parameters seems likewise to be highly intrinsically unlikely—even though it is not at all unexpected given that such parameters are observed. Do not highly intrinsically unlikely events call out for explanation, of the kind offered by the design hypothesis? While this reasoning still needs to be made precise in various ways, the outline of the fine-tuning argument is clear enough, and it deserves to be taken seriously.

One way of resisting any probability boost to the design hypothesis is to maintain that the occurrence of fine-tuned parameters is not an intrinsically unlikely outcome. This is where multiverses have often entered the story. Positing the right kind of multiverse, one which includes a universe for every possible combination of constant values, seems to achieve the required result: it is not intrinsically unlikely that somewhere in that multiverse there is a universe with a life-permitting combination of constant values.

For those disinclined towards design explanations, it has been tempting to reach for a multiverse in response to fine-tuning, and regard fine-tuning itself as the evidence for the multiverse. However, a well-known objection to the appeal to a multiverse, put forward by Hacking (1987) and White (2000), is that it still seems unlikely that this very universe—the one we in fact inhabit—has fine-tuned constant values. After all, most universes in the multiverse do not. As responsible epistemic agents, we know to take into account the whole of our evidence—and our evidence tells us that this universe is fine-tuned, not merely that some universe is fine-tuned. This logically stronger evidence seems no less unlikely given a multiverse cosmology than it is given a single universe cosmology. This particular universe, we are tempted to reason, had only a minute chance of ending up with the right parameters—so you and I had only a minute chance of existing. At least on the assumption that the universes are causally isolated, the existence or non-existence of lots of other universes doesn't seem to make any difference to the probability that this universe is fine-tuned. Then the existence of a multiverse doesn't make the fine-tuning of our universe more probable, and the evidence that our universe is fine-tuned does not support the multiverse hypothesis over the single-universe hypothesis. This argument has I think been influential in undermining the credibility of 'multiverse responses' to the fine-tuning argument within recent epistemology.

There is currently no consensus on these matters; authors including Bostrom (2002) and Bradley (2012) have offered a variety of responses to White and Hacking, and Hawthorne & Isaacs (2018) respond to a number of other criticisms of the fine-tuning argument. Where multiverses have entered the picture, the focus of the debate has tended to be on whether someone not committed to the existence of a multiverse should regard fine-tuning evidence as supporting the hypothesis of a multiverse. In section 6.3 I want to focus on a different question: if we take ourselves to have evidence for some multiverse theory on independent physical grounds, then how should we think about the epistemic import of fine-tuning evidence?

6.3 Multiverse Hypotheses as Undercutting Defeaters

As far as I know, the question of how fine-tuning reasoning is affected by the supposition that a multiverse exists has not been much explored. One exception is Roger White who writes, apparently as something of an afterthought to his defence of the objection referenced above:

> the Multiple Universe hypothesis screens off the probabilistic link between the Design hypothesis and the fine-tuning data. Hence if we happened to know, on independent grounds, that there are many universes, the

> fine-tuning facts would give us little reason to question whether the big bang was an accident, and hence our knowledge of the existence of many universes would render the fine-tuning of our universe unsurprising.
>
> White (2000: 273–4)

To put White's point another way: independent evidence for a suitable multiverse is an undercutting defeater for the design hypothesis. Such evidence does not weigh directly against the existence of a designer, as a rebutting defeater would (perhaps the problem of evil is a candidate rebutting defeater for the fine-tuning argument, given some extra premises about the likely nature of a designer?); rather, evidence from physics in favour of an appropriate multiverse is *ipso facto* higher-order evidence that fine-tuning evidence does not support the existence of a designer.

We may use the familiar analogy of misleading lighting. An object looks red to us (a fine-tuned universe looks designed) so we conclude that it is red (so we conclude there is a designer); but, once we are informed that the object is being illuminated with red light (once we are informed that there is a multiverse), we recognize that we now ought to revert to our prior expectations about the object's colour (we now ought to revert to our prior expectation about whether there is a designer). Higher-order evidence about the misleading lighting screens off the evidential relevance of our perceptual experience to the object's colour. (Information about the existence of the multiverse screens off the evidential relevance of the fine-tuning evidence to the existence of a designer.)

There are numerous interesting open questions about exactly how undercutting defeat works, raised in recent work by Lasonen-Aarnio, Sturgeon and others. Does it work by providing reasons for positive higher-order beliefs about causal or other explanatory relationships between the posited phenomenon and our possession of the evidence that seems to count in favour of that phenomenon? Or does it work purely by pruning away features of our epistemic states, without itself providing us with any positive reason for belief in any proposition? On the former model, the independent evidence that there is a multiverse provides new positive reason to believe the higher-order thesis: that the evidence that this universe is fine-tuned fails to support the thesis that there is a designer. On the latter model, the independent evidence that there is a multiverse merely cuts away some structure within our epistemic states, eliminating the link between fine-tuning and design. There is also an active debate about how undercutting defeat can and should be rendered in a Bayesian framework (see, e.g., Weisberg 2015). We need not pursue these questions here. What concerns us is which sorts of multiverses are capable of acting as undercutting defeaters and why; we can set aside the nature of undercutting defeat, and its proper representation within formal epistemology.

Before we look in more detail at the kinds of multiverses for which there might be independent evidence, and assess whether they really do act as undercutting defeaters for the evidence from fine-tuning, it is worth observing that the posit of a multiverse for reasons independent of fine-tuning reasons does not eliminate the evidential import of fine-tuning altogether. Even if evidence of fine-tuning does not support the multiverse hypothesis, and even if the multiverse hypothesis screens off the support provided by the fine-tuning evidence for the existence of a designer, the evidence of fine-tuning may still support other surprising conclusions. By analogy, your having an experience as of a red object may support some potentially surprising conclusions even if it does not support the misleading-lighting hypothesis, and even if the misleading-lighting hypothesis screens off its support for the red-object hypothesis. For example, it may support the hypothesis that the inhabitant of the room likes the colour red, or it may support the hypothesis that you can see in colour.

So: what should multiverse proponents regard as the evidential import of fine-tuning evidence? It goes without saying that the answer depends on which kind of multiverse is posited. Physicists do not multiply universes first and ask questions about what those universes are like later, even if this impression might be gained from certain philosophical work on the topic. Rather, they posit certain kinematical structures and dynamical laws in order to explain observed physical phenomena, and then ask questions about whether these physical posits give rise to multiplicities of universes. There is no general argument to be found in physics for the existence of a multiverse of some kind or other; there are only arguments for multiverses that are realized in certain physically specific ways.

6.4 Which Multiverses Can Undercut Fine-tuning?

Max Tegmark's classification of multiverses into levels (Tegmark 2003) is coarse-grained, but it provides a useful starting point:

- Level 1: Multiplicity of regions of a single spacetime, spatiotemporally distant from one another. All regions share the same physical parameters.

- Level 2: Multiplicity of regions of a single spacetime, spatiotemporally distant from one another. Regions differ in their physical parameters.
- Level 3: Multiplicity of quantum-mechanical worlds, as in EQM.
- Level 4: Multiplicity of complete possible physical realities, as in Lewisian modal realism.

Level 1 multiverses are spatially infinite universes which are ergodic in the sense that everything happens somewhere: all physically possible dynamical processes are to be found somewhere within such a universe. If you were to travel far enough within a Level 1 multiverse, you would eventually come across another region of space with indiscernible contents to our own region—for any arbitrarily large region of space around us that one may want to consider. For the limiting case of an Hubble volume indiscernible from our own, Tegmark estimates one would expect to travel $10^{10^{115}}$ metres before finding one. Still, if we live in a Level 1 multiverse, duplicates of our Hubble volume are certainly out there somewhere. Various theories of cosmic inflation seem to predict a Level 1 multiverse; but the details will not matter for our purposes. This is because knowledge of the existence of a Level 1 multiverse would not, after all, screen off the evidential relevance of a fine-tuned universe to the existence of a designer.

Why not? Because all regions in a Level 1 multiverse have the same values of the parameters that are at issue in the fine-tuning argument. Either all regions have parameter values congenial to life (even though not all of them will actually contain life, of course) or no regions do. Evidently, since we exist, if we do live in a Level 1 multiverse then we live in one in which all of the worlds have suitable parameter values for life. The existence of such a Level 1 multiverse would then seem to call out for explanation in just the same way that a single fine-tuned Hubble volume would; even if a fine-tuned multiverse is no less likely than a fine-tuned universe, it certainly does not seem any more likely. So the fine-tuning evidence remains highly surprising even on the supposition that we live in a Level 1 multiverse, and the existence of such a multiverse is not an undercutting defeater for the fine-tuning argument for a designer.

Level 2 multiverses are a different story. While they are like Level 1 multiverses in that they consist in single infinite spacetimes with different phenomena in different regions, Level 2 multiverses have different values of the parameters in different regions—and typically they are also assumed to be ergodic in our rough sense: all physically possible states of affairs—including all physically possible combinations of parameters—occur somewhere in some region of the multiverse. Hence Level 2 multiverses are capable of acting as undercutting defeaters for the support that fine-tuning evidence provides for a designer. If there is a Level 2 multiverse, then there are certain to be infinitely many different regions of spacetime that have appropriate parameter values for life, and—given that we ourselves are alive—it is no surprise that we observe a region of that kind. Whether Bradley or White is right in their assessment of the nature of the maximal relevant evidence—that this universe is suitable for life, or that we inhabit a universe suitable for life—in a Level 2 multiverse there is guaranteed to be a universe that is indiscernible from this one, so our maximal relevant evidence is guaranteed to be received somewhere. That we receive such evidence is accordingly neither surprising nor unlikely given that there exists a Level 2 multiverse.

By construction, Level 1 multiverses do not undercut the fine-tuning argument and Level 2 multiverses do. It is part of what it is to be a Level 1 multiverse that parameters do not vary across regions, and part of what it is to be a Level 2 multiverse that parameters do so vary. The non-trivial question that remains is whether we have any reason to think we live in a Level 2 multiverse, and hence any reason to think that the fine-tuning argument really is undercut. While the theories of cosmic inflation that lead to a Level 1 multiverse are relatively mainstream, the theories that generate Level 2 multiverses are much more speculative. A variety of mechanisms for generating such multiverses have been considered—for example, Linde's chaotic inflation model (Linde 1986), also known as eternal inflation, and Smolin's cosmological natural selection model (Smolin 1997) —but each proposed mechanism goes well beyond the orthodox Λ-CDM cosmology that is currently favoured by most cosmologists.

It is safe to say that is an open theoretical question whether there is a Level 2 multiverse. Our situation is thus like that someone who has seen an object that looks red, but who has also been warned that misleading lighting is a live possibility. On the supposition that there is misleading lighting, the evidential support of the red appearances for the thesis that the object is red is undercut; on the supposition that there is not misleading lighting, that evidential support is not undercut. In such a circumstance it is typically rational to reduce one's confidence that the object is in fact red below the level of confidence usually associated with red appearances when no suspicions have been raised, but to maintain that confidence above one's baseline expectation that the object is red prior to any observation of it whatever. Likewise, the evidence of fine-tuning ought to raise our confidence that there is a designer above the baseline, but this confidence ought to stay below the level that the fine-tuning evidence would establish in the absence of any suspicions of a multiverse. The more confidence we have in a Level 2 multiverse, the more confident we should be that the fine-tuning evidence is undercut and the closer our confidence in a designer should be to its baseline level.

6.5 Everettian Multiverses as Undercutting Defeaters

I now want to turn to the main target of my discussion: the consequences of the existence of a Level 3 multiverse for the evidential force of fine-tuning evidence. The Level 3 multiverse is the multiverse of EQM, and it contains an Everett world for every physically possible course of events. Unlike Level 1 and Level 2 multiverses, the universes of the Everettian multiverse are not different regions within a single infinite spacetime. If there is at least one Level 2 multiverse and in addition EQM is correct, then there is a huge plurality of Level 2 multiverses: each Everett world contains its own Level 2 multiverse.

In chapters 2 and 3 I defended a diverging version of EQM on grounds related to the interpretation of objective probability. The distinction between overlap and divergence is largely orthogonal to our present concern, however. This is because which qualitative possibilities are realized in the Everettian multiverse does not depend on how these qualitative possibilities are mereologically structured. Whether divergence or overlap is correct, there either are Everett worlds containing a variety of combinations of parameter values or there are not. If there are such worlds, then the Everettian multiverse undercuts the evidential import of fine-tuning for a designer. If there are not such worlds, then the Everettian multiverse does not undercut the fine-tuning argument.

An Everettian multiverse's ability to undercut the fine-tuning argument depends only on the existence of worlds in it with appropriate parameter values; it doesn't matter for our present purposes of assessing the fine-tuning argument how these worlds are mereologically related. What does matter for our purposes, however, is whether the worlds of a Level 3 multiverse include worlds in which there are a suitable variety of combinations of parameter values to make it unsurprising that there are life-permitting combinations. It is characteristic of Everettian multiverses that they include worlds corresponding to all physically possible outcomes of indeterministic quantum-mechanical processes. That is, if there is a non-zero quantum-mechanical chance of some outcome—no matter how small—then there is an Everett world in which that outcome occurs. Hence our question becomes: is there a quantum-mechanical chance, no matter how small, of the parameter values taking all of the combinations needed to make the existence of a world with life-permitting parameter values unsurprising? Are there indeterministic dynamical processes that assign non-zero quantum-mechanical chances both to combinations of parameter values that are life-permitting and to combinations that are not, such that the overall range of parameters permitted is of a kind that is not suggestive of design?

A toy example may help clarify matters. Suppose that only one parameter is involved—call it Z—and suppose that Z may take any integer value from 1 to 100. Only a Z value of 77 is compatible with life. A Z value of 77 is observed. *Prima facie*, this whole body of evidence tends in the context of a single-universe cosmology to support the hypothesis of a designer who selected 77 as the value for Z. Now suppose that EQM is correct, and that there exists a quantum-mechanically chancy process which determines the value of Z. There will then be Everett worlds with each of the physically possible values of Z. Now consider four different hypotheses about the chancy process which fixes the value of Z:

- Process A: The quantum probability of Z taking value 4 is 50 per cent, and the quantum probability of Z taking value 77 is 50 per cent. All other values get zero probability.
- Process B: The quantum probability of Z taking value n is 0.01 per cent for each integer n from 1 to 100 except for n=77; the quantum probability of Z taking value 77 is 99.01 per cent.
- Process C: The quantum probability of Z taking value n is 1 per cent for each integer n from 1 to 100.
- Process D: The quantum probability of Z taking value n is (n/50.5) per cent for each integer n from 1 to 100.

Which of these processes gives rise to an Everettian multiverse capable of undercutting the toy fine-tuning argument based on the value of Z?

Process A does not give rise to a suitable multiverse. Even though it guarantees that there will be an Everett world with a life-conducive value of Z, this is not enough to undercut the support that is provided for the designer hypothesis. This is because life-conducive parameter values continue to play an unexplained and unexpected role in the theory. On the supposition that two specific values of n play an unexplained and basic role in the theory, it remains very unlikely that 77 will be one of these values, and hence that life will be possible at all in our toy multiverse; given an even prior probability distribution over which pair of Z values are physically possible, the probability of one of these values being 77 is only 2 per cent. So there would still be a significant boost in this toy scenario for the design hypothesis.

Process B also does not give rise to a suitable multiverse. Although all values of Z are now rendered physically possible, so there will be an Everett world with each of the values, there is still something distinguished and special about the life-supporting value of Z: it is nearly 1,000 times more likely than any other value of Z, and there is no explanation for this fact from within the theory. So there would still be a significant boost in this toy scenario for the design hypothesis.

Process C does give rise to a suitable multiverse. The particular Z value that is conducive to life does not play any special role in the theory; it is not distinguished in any way from the other parameter values, so there is no basis for the hypothesis that a designer had any hand in so distinguishing it. If we were informed that Process C was part of the physics of our toy multiverse, then the toy argument for a designer from the apparent fine-tuning of Z would be undercut.

Process D also does give rise to a suitable multiverse. As with Process C, the particular Z value that is conducive to life does not play any special role in the theory. The probability distribution over Z values may not be uniform, but nor is it tilted in particular towards life-conducive Z values. What makes a universe more likely to be the outcome of the initial chance process, in this scenario, is just higher Z value. It is true that the life-supporting Z value is towards the higher end of the spectrum, but—as far as I can see—this fact by itself provides no significant boost to the designer hypothesis.

Note that given Processes A and B, the Everettian multiverse does not fail to undercut the designer hypothesis because a designer is needed to explain why the actual world we observe has suitable parameters. Rather, Processes A and B seem to invite the hypothesis of a designer to explain why the theory itself has certain properties that are correlated with life-conduciveness. The probabilification of a designer is not based on the observed parameter values being unlikely except if there is a designer, but instead is based on the way in which these observed parameter values are selected by an underlying causal mechanism being unlikely except if there is a designer. That alters the nature of the fine-tuning argument, but it does not change the ultimate upshot: a probability boost for the design hypothesis.

So: which of these types of scenario is actual, if EQM is correct and we are in fact living in an Everettian multiverse? The somewhat deflationary provisional conclusion of this chapter is that it is simply too early to tell. We do not know enough about the physics of the very early universe to know whether there were any dynamical processes relevant to the fixing of parameter values in the early universe. However, there is potential for progress over the coming decades in quantum gravity research to shed some light on these questions.

Candidate approaches to quantum gravity do already include appropriate candidate dynamical processes. In particular, as discussed in section 4.7, the landscape model emerging from recent work on string theory provides a mechanism by which an unstable high-dimensional spacetime state evolves into one of a staggeringly large number of different compactifications, each corresponding to a lower-dimensional spacetime characterized by a different combination of parameters. This evolution is a unitary quantum process, so there is guaranteed to be an Everett world (with its attached objective chance) that corresponds to each of the possible compactifications. And in each of these minima of the string landscape, an enormous multiplicity of parallel worlds will witness all of the different physically possible processes that play out in each of the resulting compactified spacetimes. The string landscape multiverse would make our obtaining fine-tuning evidence entirely unsurprising. Likewise, in other approaches to quantum gravity, it may reasonably be expected that some cosmological parameters may have their values dynamically determined; time will tell.

Fortunately, we can draw some epistemic lessons from the preceding discussion even in the absence of a well-confirmed theory of quantum gravity. If neither the activity of a dynamical process of the Process C/Process D sort, nor the design of a designer, was responsible for the actual parameter values, then our evidence of fine-tuning is extremely surprising even on the assumption that EQM is correct. So on the assumption that there is an Everettian multiverse, and taking into account the fine-tuned parameter values that are actually observed, there is strong support for the disjunctive hypothesis that either a life-neutral dynamical process akin to Process C and D fixed the values of the parameters in our own (region of our) Everett world or a designer was involved in setting the distribution of parameter values across the whole Everett multiverse.

Although I have argued that Everettians ought to be very confident in the above disjunction, it remains open for other doxastic commitments to tip the balance of likelihood towards one or other of these disjuncts. For example, an Everettian with very low prior credence in the existence of a designer is likely to become strongly confident in the parameter-fixing-dynamical-process disjunct, while an Everettian with prior theistic commitments is likely to become more confident in the designer disjunct. But this differential response remains well within the bounds of reasonable disagreement.

6.6 Summary

Close attention to specific fundamental cosmological hypotheses, and in particular to candidate dynamical processes that might give rise to variation in parameter value, is necessary to settle the status of the fine-tuning argument. The fine-tuning argument might be undercut by future cosmological discoveries in two main ways. Either future physics may unearth evidence of a Level 2 multiverse, or future physics may unearth evidence of life-neutral dynamical processes that operate to fix parameter values and, in conjunction with EQM, generate a Level 3 multiverse with different parameter values in different Everett worlds.

EQM, while not itself undercutting the fine-tuning argument, does nevertheless provide a cosmological framework suitable to host dynamical processes by which the fine-tuning argument might be undercut. This potential route to undercutting the fine-tuning argument is distinct from (though compatible with) to the route to undercutting the fine-tuning argument that goes via a Level 2 multiverse. A suitable dynamical parameter-fixing process need not give rise to a Level 2 multiverse in order to undercut the fine-tuning argument—though it might well give rise to one, for example if the string landscape hypothesis is combined with eternal inflation. We may conclude that there is at least one additional route to undercutting the fine-tuning argument that is available to Everettians but not to non-Everettians. Perhaps surprisingly, then, choice between interpretations of quantum mechanics turns out to be indirectly evidentially relevant to the existence of a cosmic designer.

Conclusion

An Expanding Reality

Quantum modal realists are heirs to something very like David Lewis's 'philosopher's paradise' of possibilia—except theirs is a streamlined and naturalized version of paradise. Quantum modal realism identifies compelling candidates to play the possible world role and the chance role, thereby accounting for the difference between contingency and non-contingency and enabling schematic explanations of the function of our modal thought and language. The theoretical resources of quantum modal realism, comprising a space of Everett worlds supplemented by a global chance measure, include all manner of weird and wonderful possibilia and they allow us to construct simple but powerful reductive theories of laws, counterfactuals, causation, content, and properties. Quantum modal realism provides us with all we need to move beyond objective modality and explore the outer reaches of modal thought.

The reality encompassed by quantum modal realism is certainly very large. It includes not only everything that is actual, but everything that could have been actual. The whole of reality as it is acknowledged by typical one-world quantum theories maps on to just one out of the innumerable Everett worlds. There is no getting away from it: quantum modal realism expands our reality beyond anything which evolution has equipped us to comprehend. That is not to say that quantum modal realism is the *most* ontologically inclusive system on the table; Lewisian modal realism definitely incorporates more worlds than there are in the Everettian multiverse, no matter what surprises cosmology may have in store for us. Still, quantum modal realism is a worldview expansive enough to raise doubts in all but the most enthusiastic readers. Such doubts, while extremely natural, should be reflected upon and then resisted. I maintain with Lewis that while qualitative parsimony is of great significance, quantitative parsimony is relatively insignificant, especially with respect to derivative aspects of reality. What matters is overall explanatory power and economy of theory, and quantum modal

The Nature of Contingency: Quantum Physics as Modal Realism. Alastair Wilson, Oxford University Press (2020).

DOI: 10.1093/oso/9780198846215.001.0001

realism is overall more explanatory and economical than any other synthesis of physics with metaphysics of modality.

Lewis complained that his modal realism was often met not by counter-arguments, but by an incredulous stare. The same stare may be levelled at quantum modal realism. I admit that quantum modal realism is not how we might pre-theoretically have expected a theory of modality to look. But pre-theoretical expectations bear little weight when it comes to very general questions of fundamental metaphysics. What matters is that a theory of modality should offer an adequate systematic treatment of our modal discourse and all that depends on it. It must preserve the truth of enough of modal discourse to avoid the charge of changing the subject, it must account for why we engage in modal discourse and how we acquire our modal knowledge, and it must harmonize with our best theories in fundamental physics. Unlike any extant theory of modality, quantum modal realism meets all of these desiderata.

Reality has consistently turned out to be larger than we previously imagined. Ancient humans speculated on the potential existence of other lands far from their own, beyond the sea. Around 250BCE Archimedes estimated the size of the universe at the equivalent of a sizeable two light years, but he was well ahead of his time;[1] most cosmologies over the next thousand years remained considerably smaller. It took until the late seventeenth century to properly measure the distance to the Sun and to the planets. The scale of the Milky Way galaxy was not appreciated until the nineteenth century; the gigantic distances between galaxies became known only after Hubble's observations in the 1920s. More recently, theories of cosmic inflation have transformed our understanding of spatial scales once again; an eternally inflating cosmos spawns endless bubble universes each with its own infinite FRLW spacetime. At every stage of the expansion of our vision, the length scales involved have increased by a factor too large for us to intuitively grasp. Quantum modal realism expands reality one level further.

There is one sense in which quantum modal realism does not expand our horizons: it does not widen the range of possibilities we recognize. Everyone but a fatalist recognizes that things could have gone differently, hence that there are non-actual possibilities. Quantum modal realism does not supplement these possibilities; instead it reimagines them. Rather than being abstract, ghostly, shadows of nature, alternative quantum possibilities are real and physical.

[1] The estimate required Archimedes to develop a whole new system for referring to very large numbers. His goal was to assess the maximum amount of sand that could exist. See Archimedes (*c.*250BCE).

References

Adams, R. M. (1974). 'Theories of Actuality', *Noûs* 8: 211–31.

Arndt, M. et al. (1999). 'Wave-Particle Duality of C_{60} Molecules', *Nature* 401(6754): 680–2.

Albert, D. (2000). *Time and Chance*. Cambridge: Harvard University Press.

Albert, D. (2010). 'Probability in the Everett Picture', in S. Saunders, J. Barrett, A. Kent, and D. Wallace (eds.), *Many Worlds? Everett, Quantum Theory, and Reality*. Oxford: Oxford University Press.

Albert, D. & Loewer, B. (1988). 'Interpreting the Many Worlds Interpretation', *Synthèse* 77: 195–213.

Alexander, S. (1920). *Space, Time, and Deity: the Gifford lectures at Glasgow, 1916–1918*. London: Macmillan.

Archimedes (c.250BCE). *The Sand Reckoner*.

Armstrong, D. (1978). *A Theory of Universals*. Cambridge: Cambridge University Press.

Armstrong, D. (1983). *What is a Law of Nature?* Cambridge: Cambridge University Press.

Armstrong, D. (1989). *A Combinatorial Theory of Possibility*. Cambridge: Cambridge University Press.

Armstrong, D. (2010). *Sketch for a Systematic Metaphysics*. Oxford: Oxford University Press.

Arntzenius, F. & Hall, E. (2003). 'On What We Know about Chance', *The British Journal for the Philosophy of Science* 54(2): 171–9.

Barnes, E. & Williams, J. R. G. (2011). 'A Theory of Metaphysical Indeterminacy', in K. Bennett & D. W. Zimmerman (eds.), *Oxford Studies in Metaphysics Volume 6*. Oxford: Oxford University Press.

Barnes, L. A. (2012). 'The Fine Tuning of the Universe for Intelligent Life', *Publications of the Astronomical Society of Australia* 29(4): 529–64.

Barnes, L. A. (forthcoming). '"The Fine-Tuning of the Universe for Life', in E. Knox & A. Wilson (eds.), *The Routledge Companion to the Philosophy of Physics*. New York: Routledge.

Barrett, J. (1999). *The Quantum Mechanics of Minds and Worlds*. Oxford: Oxford University Press.

Bealer, G. (1996). 'A Priori Knowledge and the Scope of Philosophy', *Philosophical Studies* 81: 121–42.

Beller, M. (1999). *Quantum Dialogue: The Making of a Revolution*. Chicago: University of Chicago Press.

Belnap, N. & Muller, T. (2010). 'Branching with Uncertain Semantics: Discussion Note on Saunders and Wallace, "Branching and Uncertainty"', *The British Journal for the Philosophy of Science* 61(3): 681–96.

Belnap, N., Perloff, M. & Xu, M. (2001). *Facing the Future: Agents and Choices in Our Indeterminist World.* New York: Oxford University Press.

Bennett, K. (2009). 'Composition, co-location, and meta-ontology', in D. Chalmers, D. Manley, & R. Wasserman (eds.), *Metametaphysics.* Oxford: Oxford University Press.

Bennett, K. (2017). *Making Things Up.* Oxford: Oxford University Press.

Berto, F. (2010). 'Impossible worlds and propositions: Against the parity thesis', *Philosophical Quarterly* 60(240): 471–86.

Berto, F. & Jago, M. (2019). *Impossible Worlds.* Oxford: Oxford University Press.

Bigelow, J. (1988). *The Reality of Numbers: A Physicalist's Philosophy of Mathematics.* Oxford: Clarendon Press.

Bigelow, J. & Pargetter, R. (1987). 'Beyond the Blank Stare', *Theoria* 53(2–3): 97–114.

Bird, A. (2001). 'Necessarily, Salt Dissolves in Water', *Analysis* 61(4): 267–74.

Bird, A. (2004). 'Strong Necessitarianism: The Nomological Identity of Possible Worlds', *Ratio* 17(3): 256–76.

Bird, A. (2007). *Nature's Metaphysics: Laws and properties.* Oxford: Oxford University Press.

Blackburn, S. (1986). 'Morals and Modals', in G. Macdonald & C. Wright (eds.), *Fact, Science and Morality.* Oxford: Blackwell.

Bohm, D. (1952a). 'A Suggested Interpretation of the Quantum Theory in Terms of Hidden Variables: Part 1', *Physical Review* 85(2): 166–79.

Bohm, D. (1952b). 'A Suggested Interpretation of the Quantum Theory in Terms of Hidden Variables: Part 2', *Physical Review* 85(2): 180–93.

Borghini, A. & Williams, N. (2008). 'A Dispositional Theory of Possibility', *dialectica* 62(1): 21–41.

Bostrom, N. (2002). *Anthropic Bias: Observation Selection Effects in Science and Philosophy.* New York: Routledge.

Bousso, R. (2002). 'The Holographic Principle', *Reviews of Modern Physics* 74(3): 825–74.

Bousso, R. & Polchinski, J. (2000). 'Quantization of Four-form Fluxes and Dynamical Neutralization of the Cosmological Constant', *Journal of High Energy Physics* 2000, JHEP06.

Braddon-Mitchell, D. & Nola, R. (eds.) (2009). *Conceptual Analysis and Philosophical Naturalism.* Boston: MIT Press.

Bradley, D. (2005). 'No Doomsday Argument without Knowledge of Birth Rank: a Defense of Bostrom', *Synthèse* 144(1): 91–100.

Bradley, D. (2011). 'Confirmation in a Branching World: The Everett Interpretation and Sleeping Beauty', *The British Journal for the Philosophy of Science* 62(2): 323–42.

Bradley, D. (2012). 'Four Problems About Self-Locating Belief', *Philosophical Review* 121(2): 149–77.

Bradley, D. & Fitelson, B. (2003). 'Monty Hall, Doomsday, and Confirmation', *Analysis* 63(1): 23–31.

Braithwaite, R. B. (1957). 'On Unknown Probabilities', in S. Körner (ed.), *Observation and Interpretation.* London: Butterworth.

Bricker, P. (2001). 'Island Universes and the Analysis of Modality', in G. Preyer & F. Siebeldt (eds.), *Reality and Humean Supervenience: Essays on the Philosophy of David Lewis*. Rowman & Littlefield.

De Broglie, L. (1927). 'Rapport au 5e Conseil de Physique Solvay, Brussels', Paris: Institut International de Physique Solvay.

Butterfield, J. (2006). 'Against Pointillisme about Mechanics', *The British Journal for the Philosophy of Science* 57(4): 709–53.

Callender, C. & Cohen, J. (2009). 'A Better Best System Account of Lawhood', *Philosophical Studies* 145(1): 1–34.

Cameron, R. (2007a). 'Lewisian Realism: Methodology, Epistemology, and Circularity', *Synthèse* 156(1): 143–59.

Cameron, R. (2007b). 'The Contingency of Composition', *Philosophical Studies* 136 (1): 99–121.

Cameron, R. (Forthcoming). 'Modal Conventionalism', in O. Bueno & S. Shalkowski (eds.), *The Routledge Handbook of Modality*. London: Routledge.

Cappelen, H. & Hawthorne, J. (2009). *Relativism and Monadic Truth*. Oxford: Oxford University Press.

Carnap, R. (1946). 'Modalities and quantification', *Journal of Symbolic Logic* 11: 33–64.

Carter, B. (1983). 'The anthropic principle and its implications for biological evolution', *Philosophical Transactions of the Royal Society of London* A310: 347–63.

Chalmers, D. (1996). *The Conscious Mind: In Search of a Fundamental Theory*. New York: Oxford University Press.

Chalmers, D. (2002). 'Does Conceivability Entail Possibility?', in J. Hawthorne & T. Gendler (eds.), *Conceivability and Possibility*. New York: Oxford University Press.

Chalmers, D. (2011). 'The Nature of Epistemic Space', in A. Egan & B. Weatherson (eds.), *Epistemic Modality*. Oxford: Oxford University Press.

Chalmers, D. (2012). *Constructing the World*. Oxford: Oxford University Press.

Chalmers, D. & Jackson, F. (2001). 'Conceptual Analysis and Reductive Explanation', *Philosophical Review* 110: 315–61.

Chen, Eddy. K. (forthcoming). "Realism about the Wave Function", *Philosophy Compass*. Available at https://arxiv.org/pdf/1810.07010.pdf.

Conroy, C. (2012). 'The relative facts interpretation and Everett's note added in proof', *Studies in History and Philosophy of Science Part B: Studies in History and Philosophy of Modern Physics* 43(2): 112–20.

Conroy, C. (2018). 'Everettian Actualism', *Studies in the History and Philosophy of Modern Physics* 63: 24–33.

Cresswell, M. (2005). 'Even modal realists should do the best they can', *Logique et Analyse* 189–92: 3–13.

Crull, E. (Forthcoming). 'Quantum Decoherence', in E. Knox and A. Wilson (eds.), *The Routledge Companion to the Philosophy of Physics*. New York: Routledge.

Cushing, J. T. (1994). *Quantum Mechanics: Historical Contingency and the Copenhagen Hegemony*. Chicago: University of Chicago Press.

Darby, G. (2010). 'Quantum Mechanics and Metaphysical Indeterminacy', *Australasian Journal of Philosophy* 88(2): 227–45.

Davidson, D. (1973). 'Radical Interpretation', *Dialectica* 27: 313–28.
Dennett, D. (1969). *Content and Consciousness*. London: Routledge & Kegan Paul.
Descartes, R. (1641). *Meditationes de prima philosophia, in qua Dei existentia et animae immortalitas demonstrantur*. Paris: Michel Soly.
Deutsch, D. (1985). 'Quantum Theory as a Universal Physical Theory', *International Journal for Theoretical Physics* 24: 1–41.
Deutsch, D. (1999). 'Quantum Theory of Probability and Decisions', *Proceedings of the Royal Society of London* A455: 3129–37.
DeWitt, B. (1968). 'The Everett–Wheeler interpretation of quantum mechanics', in C. DeWitt & J.Wheeler (eds.), *Battelle Rencontres: 1967 Lectures in Mathematics and Physics*. New York: W. A. Benjamin.
DeWitt, B. (1970). 'Quantum mechanics and reality', *Physics Today* 23(9).
Divers, J. (1999). 'A Genuine Realist Theory of Advanced Modalizing', *Mind* 108 (430): 217–40.
Divers, J. (2002). *Possible Worlds*. London: Routledge.
Divers, J. (2004). 'Agnosticism about other worlds: A new antirealist programme in modality', *Philosophy and Phenomenological Research* 69(3): 660–85.
Divers, J. (2010). 'On the Significance of the Question of the Function of Modal Judgment', in Hale & Hoffman (eds.), *Modality*. Oxford: Oxford University Press.
Divers, J. & Melia, J. (2002). 'The Analytic Limit of Genuine Modal Realism', *Mind* 111(441): 15–36.
Dorr, C. (2010). 'Of Numbers and Electrons', *Proceedings of the Aristotelian Society* 110(2): 133–81.
Dorr, C. (2016). 'To Be F Is To Be G', *Philosophical Perspectives* 30(1): 39–134.
Dorr, C. (MS). 'How to be a Modal Realist'. Available at https://philpapers.org/archive/DORHTB.pdf.
Douglas, M. (2003). 'The statistics of string/M theory vacua', *Journal of High Energy Physics* 2003, JHEP05.
Duhem, P. (1906). *La Théorie Physique. Son Objet, sa Structure*. Paris: Chevalier & Riviére.
Eddington, A. (1928). *The Nature of the Physical World*. Cambridge: Cambridge University Press.
Edgington, D. (2004). 'Two Kinds of Possibility', *Proceedings of the Aristotelian Society Supplementary Volume* 78: 1–22.
Egan, A. & Weatherson, B. (eds.) (2011). *Epistemic Modality*. Oxford: Oxford University Press.
Elga, A. (2000). 'Self-locating belief and the Sleeping Beauty problem', *Analysis* 60: 143–7.
Elga, A. (2001). 'Statistical Mechanics and the Asymmetry of Counterfactual Dependence', *Philosophy of Science (suppl. vol. 68, PSA 2000)*: 313–24.
Elga, A. (2004). 'Defeating Dr. Evil with self-locating belief', *Philosophy and Phenomenological Research* 69(2): 383–96.
Ellis, B. (2001). *Scientific Essentialism*. Cambridge: Cambridge University Press.
Evans, G. (1978). 'Can There Be Vague Objects?', *Analysis* 38: 208.
Everett, H. (1957a). 'The Theory of the Universal Wavefunction', Doctoral dissertation, Princeton University. Reprinted in B. DeWitt and N. Graham (eds.), *The*

Many-Worlds Interpretation of Quantum Mechanics (1973). Princeton: Princeton University Press.

Everett, H. (1957b). Letter to Bryce DeWitt, May 31. Available online at: http://www.pbs.org/wgbh/nova/manyworlds/orig-02a.html. Forthcoming in the UCISpace Hugh Everett III Archive at UC Irvine. http://ucispace.lib.uci.edu/

Fales, E. (1993). 'Are Causal Laws Contingent?', in J. Bacon, K. Campbell, & L. Reinhardt (eds.), *Ontology, Causality and Mind: Essays in Honour of D M Armstrong*. Cambridge: Cambridge University Press.

Field, H. (1989). 'Realism, Mathematics and Modality', in *Realism, Mathematics and Modality*. Oxford: Blackwell.

Fine, K. (1975). 'Vagueness, Truth and Logic', *Synthèse* 30: 265–300.

Fine, K. (2012a). 'A Difficulty for the Possible Worlds Analysis of Counterfactuals', *Synthèse* 189(1): 29–57.

Fine, K. (2012b). 'Guide to Ground', in F. Correia & B. Schnieder (eds.), *Metaphysical Grounding*. Cambridge: Cambridge University Press.

de Finetti, B. (1931). 'Sul significato soggettivo della probabilita', *Fundamenta Mathematicae* 17: 298–329.

Fodor, J. A. (1974). 'Special Sciences (Or: The Disunity of Science as a Working Hypothesis)', *Synthèse* 28: 97–115.

French, S. (2014). *The Structure of the World*. Oxford: Oxford University Press.

Friedman, M. (1974). 'Explanation and scientific understanding', *Journal of Philosophy* 71: 5–19.

Fritz, P. (2017). 'A Purely Recombinatorial Puzzle', *Noûs* 51(3): 547–64.

Gell-Mann, M. & Hartle, J. B. (1990). 'Quantum Mechanics in the Light of Quantum Cosmology', in W. Zurek (ed.), *Complexity, Entropy and the Physics of Information*. Reading: Addison-Wesley.

Gell-Mann, M. & Hartle, J. B. (1993). 'Classical Equations for Quantum Systems', *Physical Review D* 47(8): 3345.

Gott, J. R. (1993). 'Implications of the Copernican Principle for our Future Prospects', *Nature* 363: 315–9.

Ghirardi, G., Rimini, A., & Weber, T. (1986). 'Unified dynamics for microscopic and macroscopic systems', *Physical Review D* 34: 470–91.

Giustina et al. (2015). 'Significant Loophole-Free Test of Bell's Theorem with Entangled Photons', *Physical Review Letters* 115, 250401.

Glynn, L. (2010). 'Deterministic Chance', *The British Journal for the Philosophy of Science* 61(1): 51–80.

Goodman, J. (2004). 'An Extended Lewis/Stalnaker Semantics and the New Problem of Counterpossibles', *Philosophical Papers* 33(1): 35–66.

Graham, N. (1973). 'The Measurement of Relative Frequency', in B. DeWitt and N. Graham (eds.), *The Many-Worlds Interpretation of Quantum Mechanics*. Princeton: Princeton University Press.

Greaves, H. (2004). 'Understanding Deutsch's Probability in a Deterministic Multiverse', *Studies in History and Philosophy of Modern Physics* B35(3): 423–56.

Greaves, H. (2007a). 'On the Everettian Epistemic Problem', *Studies in History and Philosophy of Modern Physics* 38(1):120–52.

Greaves, H. (2007b). 'Probability in the Everett Interpretation', *Philosophy Compass* 2(1): 109–28.

Greaves, H. & Myrvold, W. (2010). 'Everett and Evidence', in S. Saunders, J. Barrett, A. Kent, & D. Wallace (eds.), *Many Worlds? Everett, Quantum Theory, and Reality*. Oxford: Oxford University Press.

Hacking, Ian (1987). 'The Inverse Gambler's Fallacy: the Argument from Design. The Anthropic Principle Applied to Wheeler Universes', *Mind* 96(383): 331–40.

Hájek, A. (MS). *Most Counterfactuals Are False*. https://web.archive.org/web/20130421051313/http://philosophy.cass.anu.edu.au/sites/default/files/Most%20counterfactuals%20are%20false.1.11.11_0.pdf (accessed 28/07/2019).

Hale, R. (2002). 'The Source of Necessity', *Philosophical Perspectives* 16: 299–319.

Hall, E. (2011). 'Review of *Laws and Lawmakers*, by Marc Lange', *Notre Dame Philosophical Reviews*, 2011.09.27

Handfield, T. & Wilson, A. (2014). 'Chance and Context', in A. Wilson (ed.), *Chance and Temporal Asymmetry*. Oxford: Oxford University Press.

Hawthorne, J. (2005). 'Chance and Counterfactuals', *Philosophy and Phenomenological Research* 70(2): 396–405.

Hawthorne, J. & Isaacs, Y. (2018). 'Fine-tuning Fine-tuning', in M. A. Benton, J. Hawthorne, & D. Rabinowitz (eds.), *Knowledge, Belief, and God: New Insights in Religious Epistemology*. Oxford: Oxford University Press.

Heller, M. (2003). 'The Immorality of Modal Realism, or: How I learned to stop worrying and let the children drown', *Philosophical Studies* 114: 1–22.

Hensen et al. (2015). 'Loophole-free Bell inequality violation using electron spins separated by 1.3 kilometres', *Nature* 526: 682–6.

Hitchcock, C. (2001). 'The Intransitivity of Causation Revealed in Equations and Graphs', *Journal of Philosophy* 98(6): 273–99.

Hoefer, C. (2007). 'The Third Way on Objective Probability: A Sceptic's Guide to Objective Chance', *Mind* 116(463): 549–96.

Hudson, H. (1997). 'Brute Facts', *Australasian Journal of Philosophy* 75(1): 77–82.

Jansson, L. (2016). 'Everettian quantum mechanics and physical probability: Against the principle of "State Supervenience"', *Studies in History and Philosophy of Modern Physics* 53: 45–53.

Jacobs, J. D. (2010). 'A Powers Theory of Modality: Or, how I Learned to Stop Worrying and Reject Possible Worlds', *Philosophical Studies* 151(2): 227–48.

Jago, M. (2014). *The Impossible*. Oxford: Oxford University Press.

Jackson, F. (1998). *From Metaphysics to Ethics: A Defence of Conceptual Analysis*. Oxford: Clarendon Press.

Jubien, M. (1988). 'Problems with Possible Worlds', in D. F. Austin (ed.), *Philosophical Analysis*. Dordrecht: Kluwer Academic Publishers.

Kent, A. (2010). 'One World Versus Many: The Inadequacy of Everettian Accounts of Evolution, Probability, and Confirmation', in S. Saunders, J. Barrett, A. Kent, & D. Wallace (eds.), *Many Worlds? Everett, Quantum Theory, and Reality*. Oxford: Oxford University Press.

Kim, J. (1992). 'Multiple Realization and the Metaphysics of Reduction', *Philosophy and Phenomenological Research* 52(1): 1–26.

Kistler, M. (2002). 'The Causal Criterion of Reality and the Necessity of Laws of Nature', *Metaphysica* 3(1): 57–86.

Kitcher, P. (1981). 'Explanatory unification', *Philosophy of Science* 48: 507–31.

Kripke, S. (1980). *Naming & Necessity*. Cambridge: Harvard University Press.

Ladyman, J. (1998). 'What is Structural Realism?', *Studies in History and Philosophy of Science* A29(3): 409–24.

Ladyman, J. & Ross, D. (2007). *Every Thing Must Go: Metaphysics Naturalized* (with J. Collier & D. Spurrett). Oxford: Oxford University Press.

Lange, M. (2008). 'Why contingent facts cannot necessities make', *Analysis* 68(298): 120–8.

Lange, M. (2009). *Laws and Lawmakers*. New York: Oxford University Press.

Leslie, J. (1989). *Universes*. London: Routledge.

Lewis, D. K. (1968). 'Counterpart Theory and Quantified Modal Logic', *Journal of Philosophy* 65: 113–26.

Lewis, D. K. (1970). 'Anselm and Actuality', *Noûs* 4: 175–88.

Lewis, D. K. (1973). *Counterfactuals*. Oxford: B. Blackwell.

Lewis, D. K. (1974). 'Radical Interpretation', *Synthèse* 23: 331–44. Reprinted with postscripts in Lewis (1983a).

Lewis, D. K. (1979). 'Counterfactual Dependence and Time's Arrow'. *Noûs* 13: 455–76. Reprinted with postscripts in Lewis (1986a).

Lewis, D. K. (1980/1983). 'A Subjectivist's Guide to Objective Chance', in R. Jeffrey (ed.), *Studies in Inductive Logic and Probability Vol.2*. Berkeley: University of California Press. Reprinted with postscripts in Lewis (1983a).

Lewis, D. K. (1983a). *Philosophical Papers Vol. I*. New York: Oxford University Press.

Lewis, D. K. (1983b). 'New Work for a Theory of Universals', *Australasian Journal of Philosophy* 61(4):343–77.

Lewis, D. K. (1986a). *Philosophical Papers Vol. II*. New York: Oxford University Press.

Lewis, D. K. (1986b). *On the Plurality of Worlds*. Oxford: B. Blackwell.

Lewis, D. K. (1988a). 'Relevant Implication', *Theoria* 54(3): 161–74.

Lewis, D. K. (1988b). 'Statements Partly About Observation', *Philosophical Papers* 17(1): 1–31.

Lewis, D. K. (1994). 'Chance and Credence: Humean Supervenience Debugged', *Mind* 103(412): 473–90.

Lewis, D. K. (1996). 'Maudlin and Modal Mystery', *Australasian Journal of Philosophy* 74(4): 683–4.

Lewis, D. K. (2004). 'How Many Lives Has Schrödinger's Cat?', *Australasian Journal of Philosophy* 82(1): 3–22.

Lewis, D. K. (2009). 'Ramseyan Humility', in D. Braddon-Mitchell & R. Nola (eds.), *Conceptual Analysis and Philosophical Naturalism*. Boston: MIT Press.

Lewis, P. (2009). 'Probability, Self-Location, and Quantum Branching', *Philosophy of Science* 76(5): 1009–19.

Linde, A. (1986). 'Eternally Existing Self-Reproducing Chaotic Inflationary Universe', *Physics Letters B* 175(4): 395–400.

Lockwood, M. (1989). *Mind, Brain and the Quantum: The Compound 'I'*. Cambridge: Blackwell.

Loewer, B. (1996). 'Comment on Lockwood', *The British Journal for the Philosophy of Science* 47(2): 229–32.

Marshall, D. (2016). 'A Puzzle for Modal Realism', *Philosophers' Imprint* 16(9): 1–24.

MacFarlane, J. (2014). *Assessment Sensitivity: Relative Truth and its Applications*. Oxford: Oxford University Press.

Martin, C. (2008). *The Mind in Nature*. New York: Oxford University Press.

Maudlin, T. (1994). *Quantum Non-Locality and Relativity*. Cambridge: Blackwell.

Maudlin, T. (1996). 'On the Impossibility of David Lewis's Modal Realism', *Australasian Journal of Philosophy* 74(4): 669–82.

Maudlin, T. (2007). *The Metaphysics within Physics*. Oxford: Oxford University Press.

McGee, V. & McLaughlin, B. (1995). 'Distinctions Without a Difference', *Southern Journal of Philosophy* 33: 203–51.

McLaughlin, B. & Bennett, K. (2018). 'Supervenience', *Stanford Encyclopaedia of Philosophy*. Version of Wed Jan 10, 2018. https://plato.stanford.edu/entries/supervenience/ (accessed 28/07/19).

McMullin, E. (1978). 'Structural Explanation', *American Philosophical Quarterly* 15(2): 139–47.

Meacham, C. (2005). 'Three proposals regarding a theory of chance', *Philosophical Perspectives* 19(1): 281–307.

Melia, J. (2003). *Modality*. Routledge.

Mill, J. S. (1843). *A System of Logic*. London: John W. Parker.

Monton, B. (2003). 'The Doomsday Argument without Knowledge of Birth Rank', *The Philosophical Quarterly* 53(210): 79–82.

Myrvold, W. (2002). 'On Peaceful Coexistence: Is the Collapse Postulate Incompatible with Relativity?', *Studies in History and Philosophy of Modern Physics* 33B(3): 435–66.

Nagel, E. (1961). *The Structure of Science: Problems in the Logic of Scientific Explanation*. New York: Harcourt Brace World.

Neurath, O. (1944). 'Foundations of the Social Sciences', in O. Neurath, R. Carnap, & C. Morris (eds.), *International Encyclopedia of Unified Science* 2:1. Chicago: University of Chicago Press.

Nolan, D. (1996). 'Credo'. https://sites.google.com/site/professordanielnolan/credo (accessed 31/10/19).

Nolan, D. (1997). 'Impossible Worlds: A Modest Proposal', *Notre Dame Journal of Formal Logic* 38(4): 535–72.

Nolan, D. (2010). 'Comments on John Divers' "On the Significance of the Question of the Function of Modal Judgment"', in A. Hale & A. Hoffman (eds.), *Modality*, pp. 220–6. Oxford: Oxford University Press.

Noonan, H. W. (1994). 'In Defence of the Letter of Fictionalism', *Analysis* 54(3): 133–9.

Oppenheim, P. & Putnam, H. (1958). 'The unity of science as a working hypothesis', *Minnesota Studies in the Philosophy of Science* 2: 3–36.

Papineau, D. (1996). 'Many Minds are No Worse than One', *British Journal for the Philosophy of Science* 47(2): 233–41.

Papineau, D. (2010). 'A Fair Deal for Everettians', in S. Saunders, J. Barrett, A. Kent, & D. Wallace (eds.), *Many Worlds? Everett, Quantum Theory, and Reality*. Oxford: Oxford University Press.

Parfit, D. (1984). *Reasons and Persons*. Oxford: Clarendon Press.

Pargetter, R. (1984). 'Laws and Modal Realism', *Philosophical Studies* 46: 335–48.

Parsons, J. (MS). 'Against Advanced Modalizing'. Available at https://citeseerx.ist.psu.edu/viewdoc/summary?doi=10.1.1.703.3458 (accessed 29/07/2019).

Plantinga, A. (1974). *The Nature of Necessity*. Oxford: Clarendon Press.

Price, H. (2010). 'Decisions, Decisions, Decisions: Can Savage Salvage Everettian Probability?', in S. Saunders, J. Barrett, A. Kent, & and D. Wallace (eds.), *Many Worlds? Everett, Quantum Theory, and Reality*. Oxford: Oxford University Press.

Price, H. & Jackson, F. (1997). 'Naturalism and the Fate of the M-Worlds', *Aristotelian Society: Supplementary Volume* 71: 247–82.

Priest, G. (2005). *Towards Non-Being: The Logic and Metaphysics of Intentionality*. New York: Oxford University Press.

Quine, W. V. (1951). 'Two Dogmas of Empiricism', *Philosophical Review* 60: 20–43.

Quine, W. V. (1957). 'The Scope and Language of Science', *British Journal for the Philosophy of Science* 8: 1–17.

Quine, W. V. (1960a). *Word and Object*. Cambridge: MIT Press.

Quine, W. V. (1960b). 'Carnap and Logical Truth', *Synthèse* 12: 350–74.

Quine, W. V. (1969a). 'Natural Kinds', in *Ontological Relativity and Other Essays*. New York: Columbia University Press.

Quine, W. V. (1969b). 'Propositional Objects', in *Ontological Relativity and Other Essays*. New York: Columbia University Press.

Ramsey, F. P. (1928). 'Universals of Law and of Fact', reprinted in D. H. Mellor (ed.), *F. P. Ramsey: Philosophical Papers* (1990). Cambridge: Cambridge University Press.

Rees, M. (2000). *Just Six Numbers: The Deep Forces that Shape the Universe*. New York: Basic Books.

Restall, G. (1996). 'Truthmakers, Entailment and Necessity', *Australasian Journal of Philosophy* 74: 331–40.

Restall, G. (1997). 'Ways Things Can't Be', *Notre Dame Journal of Formal Logic* 38: 583–96.

Reutlinger, A. (2017). 'Does the Counterfactual Theory of Explanation Apply to Non-Causal Explanations in Metaphysics?', *European Journal for Philosophy of Science*, 7(2): 239–56.

Richards, T. (1975). 'The Worlds of David Lewis', *Australasian Journal of Philosophy* 53: 105–18.

Rosen, G. (1990). 'Modal Fictionalism', *Mind* 99(395): 327–54.

Rosen, G. (2006). 'The Limits of Contingency', in F. MacBride (ed.), *Identity and Modality*. Oxford: Clarendon Press.

Saunders, S. (1993). 'Decoherence, relative states, and evolutionary adaptation', *Foundations of Physics* 23: 1553–85.

Saunders, S. (1994). 'What Is the Problem of Measurement?', *Harvard Review of Philosophy* 4(1): 4–22.

Saunders, S. (1995). 'Time, quantum mechanics, and decoherence', *Synthèse* 102: 235–66Saunders, S. (1997). 'Naturalizing Metaphysics', *The Monist* 80(1): 44–69.
Saunders, S. (1998). 'Time, Quantum Mechanics, and Probability'. *Synthèse* 114(3): 373–404.
Saunders, S. (2005). 'What is Probability?', in A. Elitzur, S. Dolev, & N. Kolenda (eds.), *Quo Vadis Quantum Mechanics?*. Dordrecht: Springer.
Saunders, S. (2010a). 'Many Worlds? An Introduction', in S. Saunders, J. Barrett, A. Kent, & D. Wallace (eds.), *Many Worlds? Everett, Quantum Theory, and Reality*. Oxford: Oxford University Press.
Saunders, S. (2010b). 'Chance in the Everett Interpretation', in S. Saunders, J. Barrett, A. Kent, & D. Wallace (eds.), *Many Worlds? Everett, Quantum Theory, and Reality*. Oxford: Oxford University Press.
Saunders, S. & Wallace, D. (2008). 'Branching and Uncertainty', *The British Journal for the Philosophy of Science* 59(3): 293–305.
Savage, L. J. (1972). *The Foundations of Statistics* (2nd ed.). New York: Dover Publications.
Schaffer, J. (2003). 'Principled Chances', *The British Journal for the Philosophy of Science* 54(1): 27–41.
Schaffer, J. (2005). 'Quiddistic Knowledge', *Philosophical Studies* 123(1): 1–32.
Schaffer, J. (2007). 'Deterministic Chance', *The British Journal for the Philosophy of Science* 58(2): 113–40.
Schaffer, J. (2009). 'On What Grounds What', in D. Chalmers, D. Manley, & R. Wasserman (eds.), *Metametaphysics*. Oxford: Oxford University Press.
Schaffer, J. (2010). 'Monism: The Priority of the Whole', *Philosophical Review* 119(1): 31–76.
Schaffer, J. (2014). 'What Not to Multiply Without Necessity', *Australasian Journal of Philosophy* 93(4): 644–64.
Schaffer, J. (2016). 'Grounding in the Image of Causation', *Philosophical Studies* 173 (1): 49–100.
Schaffer, J. & Ismael, J. (Forthcoming). 'Quantum holism: non-separability as common ground'. To appear in *Synthèse*.
Schrenk, M. (2006). 'A Theory for Special Science Laws', in H. Bohse & S. Walter (eds.), *Selected Papers Contributed to the Sections of Gap.6*. Mentis.
Sebens, C. & Carroll, S. (2018). 'Self-Locating Uncertainty and the Origin of Probability in Everettian Quantum Mechanics', *The British Journal for the Philosophy of Science* 69(1): 25–74.
Sellars, W. (1948). 'Concepts as Involving Laws and Inconceivable Without Them', *Philosophy of Science* 15(4): 287–315.
Sellars, W. (1957). 'Counterfactuals, Dispositions, and the Causal Modalities', in H. Feigl, M. Scriven, and G. Maxwell (eds.), *Minnesota Studies in the Philosophy of Science, Vol. II*: 225–308. Minneapolis: University of Minnesota Press.
Skow, B. (2010). 'Deep Metaphysical Indeterminacy', *Philosophical Quarterly* 60: 851–8.
Skow, B. (2016). *Reasons Why*. Oxford: Oxford University Press.
Skyrms, B. (1980). *Causal Necessity*. New Haven: Yale University Press.
Shalm, L. et al. (2015). 'Strong Loophole-Free Test of Local Realism', *Physical Review Letters* 115: 250402.

Shoemaker, S. (1980). 'Causality and Properties', in *Time and Cause: Essays Presented to Richard Taylor*. Dordrecht: Springer.
Shoemaker, S. (1998). 'Causal and Metaphysical Necessity', *Pacific Philosophical Quarterly* 79 (1): 59–77.
Sider, T. (2000). 'Reductive Theories of Modality', in M.J. Loux & D.W. Zimmerman (eds.), *The Oxford Handbook of Metaphysics*. New York: Oxford University Press.
Sider, T. (2011). *Writing the Book of the World*. Oxford: Oxford University Press.
Sider, T. (MS). 'Beyond the Humphrey Objection', http://tedsider.org/papers/counterpart_theory.pdf (accessed 31/10/19).
Smart, J. J. C. (1963). *Philosophy and Scientific Realism*. Humanities Press.
Smart, J. J. C. (1984). *Ethics, Persuasion and Truth*. Boston: Routledge & K Paul.
Smolin, L. (1997). *The Life of the Cosmos*. New York: Oxford University Press.
Sober, E. (2002). 'An Empirical Critique of Two Versions of the Doomsday Argument—Gott's Line and Leslie's Wedge', *Synthèse* 135: 415–30.
Stalnaker, R. (1968). 'A Theory of Conditionals', in N. Rescher (ed.), *Studies in Logical Theory* (American Philosophical Quarterly Monographs 2): 98–112. Oxford: Blackwell.
Stalnaker, R. (1984). *Inquiry*. Cambridge: MIT Press.
Stalnaker, R. (1996). 'Impossibilities', *Philosophical Topics* 24(1): 193–204.
Stalnaker, R. (2014). *Context*. Oxford: Oxford University Press.
Strevens, M. (1999). 'Objective Probability as a Guide to the World', *Philosophical Studies* 95(3): 243–75.
Strevens, M. (2008). *Depth: An Account of Scientific Explanation*. Cambridge: Harvard University Press.
Susskind, L. (2005). *The Cosmic Landscape: String Theory and the Illusion of Intelligent Design*. Boston: Little, Brown and Company.
Swoyer, C. (1982). 'The Nature of Natural Laws', *Australasian Journal of Philosophy* 60: 203–23.
Tappenden, P. (2008). 'Saunders and Wallace on Everett and Lewis', *The British Journal for the Philosophy of Science* 59(3): 307–14.
Tan, P. (Forthcoming). 'Counterpossible Non-vacuity in Scientific Practice', *Journal of Philosophy* 116(1): 32–60.
Tegmark, M. (2003). 'Parallel Universes', in J. D. Barrow, P. C. W. Davies, & C. L. Harper (eds.), *Science and Ultimate Reality: From Quantum to Cosmos*. Cambridge: Cambridge University Press.
Tegmark, M. & Rees, M. (1997). 'Why is the CMB fluctuation level 10^{-5}?', *Astrophysical Journal* 499: 526–32.
Tumulka, R. (2006). 'A relativistic version of the Ghirardi-Rimini-Weber model', *Journal of Statistical Physics* 125: 825–44.
Van Fraassen, B. (1980). *The Scientific Image*. Oxford: Clarendon Press.
Van Fraassen, B. (1989). *Laws and Symmetry*. New York: Oxford University Press.
Van Inwagen, P. (1986). 'Two Concepts of Possible Worlds', in French, Uehling, and Wettstein (eds.), *Midwest Studies in Philosophy* XI, pp. 185–213. Minneapolis: University of Minnesota Press.
Vetter, B. (2015). *Potentiality*. Oxford: Oxford University Press.

Wallace, D. (2003a). 'Everett and Structure', *Studies in History and Philosophy of Science* 34(1): 87–105.

Wallace, D. (2003b). 'Everettian Rationality: Defending Deutsch's Approach to Probability in the Everett Interpretation', *Studies in History and Philosophy of Modern Physics* 34(3): 415–39.

Wallace, D. (2006). 'Epistemology Quantized: Circumstances in Which We Should Come to Believe in the Everett Interpretation', *The British Journal for the Philosophy of Science* 57(4): 655–89.

Wallace, D. (2007). 'Quantum Probability from Subjective Likelihood: Improving on Deutsch's Proof of the Probability Rule', *Studies in History and Philosophy of Modern Physics* 38(2): 311–32.

Wallace, D. (2010a). 'Decoherence and Ontology', in S. Saunders, J. Barrett, A. Kent, & D. Wallace (eds.), *Many Worlds? Everett, Quantum Theory, and Reality*. Oxford: Oxford University Press.

Wallace, D. (2010b). 'How to Prove the Born Rule', in S. Saunders, J. Barrett, A. Kent, & D. Wallace (eds.), *Many Worlds? Everett, Quantum Theory, and Reality*. Oxford: Oxford University Press.

Wallace, D. (2012). *The Emergent Multiverse: Quantum Theory According to the Everett Interpretation*. Oxford: Oxford University Press.

Wallace, D. (Forthcoming). 'The Logic of the Past Hypothesis'. To appear in B. Loewer, E. Winsberg, & B. Weslake (eds.), *Time's Arrows and the Probability Structure of the World*. Cambridge: Harvard University Press.

Wallace, D. & Timpson, C. G. (2010). 'Quantum Mechanics on Spacetime I: Spacetime State Realism', *The British Journal for the Philosophy of Science* 61(4): 697–727.

Wang, J. (2016). 'Fundamentality and Modal Freedom', *Philosophical Perspectives* 30(1): 39–134.

Weinberg, S. (1989). 'The Cosmological Constant Problem', *Review of Modern Physics* 61(1).

Weisberg, J. (2015). 'Updating, Undermining, and Independence', *The British Journal for the Philosophy of Science* 66 (1): 121–59.

White, R. (2000). 'Fine-Tuning and Multiple Universes', *Noûs* 34(2): 260–76.

Williamson, T. (1996). *Vagueness*. Oxford: Blackwell.

Williamson, T. (2007). *The Philosophy of Philosophy*. Oxford: Blackwell.

Williamson, T. (2013). *Modal Logic as Metaphysics*. Oxford: Oxford University Press.

Williamson, T. (Forthcoming). 'Counterpossibles in Metaphysics', to appear in B. Armour-Garb & F. Kroon, *Philosophical Fictionalism*.

Wilson, A. (2011). 'Macroscopic Ontology in Everettian Quantum Mechanics', *Philosophical Quarterly* 61(243): 363–82.

Wilson, A. (2012). 'Everettian Quantum Mechanics Without Branching Time', *Synthèse* 188(1): 67–84.

Wilson, A. (2013a). 'Objective Probability in Everettian Quantum Mechanics', *The British Journal for the Philosophy of Science* 64(4): 709–37.

Wilson, A. (2013b). 'Schaffer on Laws of Nature', *Philosophical Studies* 164(3): 653–7.

Wilson, A. (2014). 'Everettian Confirmation and Sleeping Beauty', *The British Journal for the Philosophy of Science* 65(3): 573-598.

Wilson, A. (2017). 'The Quantum Doomsday Argument', *The British Journal for the Philosophy of Science* 68(2): 597–615.

Wilson, A. (2018a). 'Metaphysical Causation', *Noûs* 52(4): 723–51.

Wilson, A. (2018b). 'Grounding Entails Counterpossible Non-Triviality', *Philosophy and Phenomenological Research* 96(3): 716–28.

Wilson, A. (MS). 'Plenitude and Recombination'.

Wilson, J. (2013). 'A Determinable-based Account of Metaphysical Indeterminacy', *Inquiry* 56: 359–85.

Wilson, J. (2014). 'Hume's Dictum and Natural Modality: Counterfactuals', in A. Wilson (ed.), *Chance and Temporal Asymmetry*. Oxford: Oxford University Press.

Wilson, M. (1985). 'What is this Thing Called "Pain"?', *Pacific Philosophical Quarterly* 66(3/4): 227–67.

Wolff, J. (2015). 'Spin as a Determinable', *TOPOI* 34(2): 379–86.

Woodward, J. (2003). *Making Things Happen: A Theory of Causal Explanation*. Oxford: Oxford University Press.

Zurek, W. H. (1993). 'Environment-Induced Decoherence and the Transition from Quantum to Classical', *Vistas in Astronomy* 37: 185–96.

Zurek, W. H. (2002). 'Decoherence and the transition from quantum to classical—REVISITED', *Los Alamos Science* 27: 2–25.

Index

For the benefit of digital users, indexed terms that span two pages (e.g., 52-53) may, on occasion, appear on only one of those pages.